USING TELEVISION AND VIDEO IN BUSINESS

USING TELEVISION AND VIDEO IN BUSINESS

Andrew Crofts

MERCURY

First published in 1991
by Mercury Books
Gold Arrow Publications Limited
862 Garratt Lane, London SW17 0NB

Set in Baskerville by TecSet Ltd. Wallington, Surrey
Printed and bound in Great Britain by
Mackays of Chatham, PLC, Chatham, Kent

British Library Cataloguing in Publication Data

Crofts, Andrew
Using television and video in business.
I. Title
658.4

ISBN 1-85251-099-4

Contents

Preface vii

Introduction 1

1. Where Has Business Television Come From? 6
2. Who Does It? How to Choose a Production House 17
3. How Much Does It Cost? 28
4. Defining Your Audience 38
5. Private Corporate Networks 46
6. The Corporate Video 61
7. Television at Conferences 70
8. Video Conferencing 77
9. Taking Screens into the Public Arena 88
10. Television as a Selling Aid 95
11. Education and Training 99
12. Training in Communications Skills 110
13. Interactive Video 120
14. Getting it Seen 130
15. Editing and Post-Production 137

16. Sponsoring Television 144

17. The Future is Here 153

Index 159

Preface

Some aspects of the business television industry still sound like the dreams of a science fiction writer, but all of them can be bought and used today, if you know how to go about it.

If you know what your needs are, who to go to for advice and what to ask them, you can make television technology do virtually whatever you want.

If you wanted to, for instance, you could put personal video-conferencing facilities on the desks of everyone working in your company.

You could also buy them personal computers which would allow them to create their own video programmes for use as sales and presentation aids.

It would be no problem to set up a private network to reach any number of outlets, broadcasting programmes which you could have written and produced to your exact specifications.

You can also commission interactive programmes and three-dimensional animations so realistic that they virtually re-create reality to your own design.

You can build screens of almost any size, and beam pictures down to them from anywhere on earth. You can use those screens to talk to anyone else in the world at any time. You can do business with them or learn from them and obtain experiences that would never be available to you in any other way.

Even if you aren't doing any of these things yet, you may already have had experience in making corporate videos, and you may have bought and hired some of the generic video training products on the market.

Yet it is still a new industry, full of uncertainties. Many business people fear it because they see a maze of technology, jargon and potential expense ahead of them, and they don't know where to start.

Television and video are the most powerful communications tools ever invented, if you can just summon up the courage to use them effectively.

Technological developments are turning the television screen into the perfect medium for two-way communication. At the same time changes in the international laws are opening up opportunities for sponsored television, and for video publishing projects which can provide companies with new opportunities to sell and promote themselves and their products to new audiences.

Within any company, video and television can now be used as tools to support the management, marketing, selling, personnel, training, sales promotion, public relations, investor relations, publishing and customer care functions.

Before long television screens will be as integrated into daily business life as telephones and computers, and people who know how to use them effectively will gain the competitive edge.

This book takes the reader through the various different ways in which television and video can be used and misused within the business world, and attempts to predict what the future will bring.

There is no 'right way' to use television in business. It is an infinitely adaptable medium, which can be used to answer any number of communications problems in any number of ways.

My aim is to blow away some of the mystery which surrounds the world of business television, and to alert readers to the opportunities which are already there, waiting to be grasped.

Andrew Crofts

Introduction

A number of changes are coming together in the 1990s to make this the age of Business Television:

- Several powerful new technologies have emerged which make it possible to communicate electronically in ways that most of us never dreamed possible in the Eighties.

  Television has become a two-way communication medium through the use of satellite, computing and telecommunications technologies. It has become flexible enough to adapt to needs within the marketplace which could never had been served by television in its traditional forms.
- The increased globalisation of companies and industries has meant that business people are having to travel further and more often than ever before.

  There is a limit to how much travel is practical before the costs, both financial and emotional, become impossible to bear. Television provides a real alternative to the face-to-face meeting and the 'pressing of flesh'.
- Managers all over the advanced world are realising that one of the most important ingredients for business success in the future will be communication, and apart from the face-to-face meeting, nothing communicates quite as effectively as television.

  Many companies now have 'visions', 'goals' or 'mission statements', and in all of them communication is cited as crucial to future success. Many of those companies, however, are still unsure how to make their actions live up to their expectations. Television holds the key.
- Increased competition in all areas of the business world means that speed is now vital in all things. The concept of 'just-in-time' management applies as much to communications and information as it does to stock control and product deliveries

  Companies can no longer wait three years for a corporate training scheme to cascade down through a company; they need to retrain entire organisations overnight. It is impossible to do that with people alone, there has to be a technological way of delivering the messages consistently, quickly and accurately. Nothing delivers messages as fast, as consistently or as accurately as television.

- The whole world has become hungry for information and training, and they want it served up in digestible, attractive, bite-sized pieces. Nothing does that as effectively as television. It does not replace other forms of training, but it is the most powerful new tool to come into the hands of the training establishment since the printing press.

Those are the needs which the business community has. In Europe the realisation that television can help to satisfy these needs has been slow in coming, for a mixture of cultural, technical and legal reasons. But all those reasons are beginning to melt away, and there are an increasing number of voices speaking in favour of the things which business television represents. The pressure for changes in the way we communicate are coming from all sides.

THE FORCES DRIVING THE CHANGES

The impetus to increase the use of television in business is coming from a number of different sources:
The boardrooms, where senior managers want to use the technology to communicate, either through traditional corporate videos or through the video-conferencing facilities and private networks which have been facilitated by the launching of satellites around the globe. They want to cut down their travelling time and improve the standards of their communications. They want to talk directly to employees, customers and all the other groups who are important to the welfare of any organisation;
The public relations departments, who realise the power of the visual medium to put across their messages and to talk to target groups who were previously inaccessible;
The human resources departments, who want to find cost-effective new ways of communicating with their people on a regular and personal basis;
The sales departments, who can see that something has to be done to curb the escalating costs of moving people around the globe for giant conferences, and who want more effective ways of demonstrating products to potential customers;
The laboratories and workshops, where technicians and engineers want to be able to talk to colleagues in other parts of their companies, and other

parts of the globe, to compare notes, ask questions and explain what it is that they are doing. They no longer want to work in isolation, they want to use technology to broaden their own knowledge and to exchange information and experiences;

The training departments, who need to find ever more efficient ways of disseminating training and educational material to ever larger audiences, and who know that only television can achieve this with the right degrees of consistency, effectiveness and urgency;

The marketing and merchandising departments, who can see that television in the public arena can be a powerful point-of-sale aid, and know that it can play a number of other roles in the marketing mix;

The advertising departments, who can see that changes in the world of broadcast television are going to open up a host of opportunities for sponsored programming;

The finance departments, who are looking for ways to cut the costs of business travel, to make their companies more efficient, and to develop cost-effective means of communicating corporate messages;

The managers, who want to use television technology to improve their interpersonal skills and manage their time more effectively.

All these people are becoming aware of just how much television technology can help them to deal with the changing demands of their jobs, and they find the whole concept exciting. Some of them also realise that unless they learn how to manage business television they may get left behind.

Imagine a company which refused to install a telephone line, or dismissed out of hand any talk about buying a word processor to replace their secretary's trusty old typewriter. Companies that resist the march of business television could find themselves in the same position within a few years.

Televisual technology is going to become as commonplace in the business world as telephones and computers, and is already inextricably entwined with both.

WHO'S SELLING IT?

With these needs building up inside companies, it is not surprising that a number of suppliers have appeared in the market who want to sell televisual products and services to the corporate world.

There are the *production houses* who make the programmes.

There are the *telecommunications companies* who want to persuade customers to use their lines to transmit programmes.

There are the *satellite and cable companies* who want client organisations to broadcast programmes via their hardware.

There are the *computer companies* wanting to be part of the video-conferencing and interactive video industries.

There are the *makers of training films* and the makers of interactive video programmes.

There are the *manufacturers* selling and hiring out the hardware.

There is the *editing and post-production industry*.

There are the *sponsorship consultancies*, the *video publishers*, the *video conferencing companies*, the *conference organisers* and the *communications consultants*.

So many people now have a vested interest in making the business television industry into a major business that it has become unstoppable.

WHAT SHOULD YOU DO ABOUT IT?

But what if you don't know anything about any of this? What if your knowledge of television extends to what you see on the few channels which make it into your living room, and the VCR machine? How do you start to make sense of all the clamouring that is going on in the marketplace? How can you work out what you should be doing about business television? And once you have worked that out, how do you set about making it a reality?

Should you be hiring someone in-house to deal with it all?

Should you be installing a studio?

How do you choose a production company that won't rip you off?

How do you commission work from them?

How do you get people to watch your programmes?

What are the golden rules of appearing on screen?

What is an interactive video?

What is the difference between a private network and a video conference?

Is it a good idea to get a great star to front your corporate video?

What is the difference between off-line and on-line editing?

What is high definition television?

How do you avoid making the same sort of mistakes that companies made twenty years ago when the computing technologies were first becoming widespread?

In this book I want to answer these sorts of questions, and at the same time to convey some of the excitement of what is happening in the business television industry. I want to suggest to the readers what their first tentative steps into the minefield should be.

It may be that you have already had experience in one of the sectors; perhaps you have made a corporate video, hired a management training film, or staged a television link-up at a conference. You may still be completely bemused by the other sectors of the market. Some of you may feel that you have had your fingers burnt on your first encounters with the industry. I hope that this book will explain how that may have happened, and how you can stop it happening again.

When computers first came onto the market many companies rushed in and bought systems which were completely unsuitable for their needs, and they made sweeping judgements of the whole computing business as a result of their first unhappy experiences. Before going into any new area it is always good to do some research into the marketplace first. I hope this book will help readers with that research.

Chapter 1

Where Has Business Television Come From?

When television was launched in Britain it was seen as the preserve of officers, gentlemen and scholars. The BBC was staffed by Oxbridge arts graduates who knew what their audiences wanted because the audiences were the same as themselves.

With the growth of the mass audience and the advent of commercial television, the influence of the wider market came to bear, and the world of business established its first foothold through the advertising revenue which it poured into the young independent television companies. Still, however, television was an élitist medium, even the people who made the commercials were from the same backgrounds as their less well-paid colleagues at the BBC.

Commercial television was memorably labelled 'a licence to print money', and television people moved apart from others in their earning power as well as their intellectual élitism.

In the mid Sixties the BBC decided the time had come to acknowledge the existence of the real world of commerce and industry, and decided to make a series called *The Money Programme*. 'Initially there was considerable debate about the title,' says Alan Watson, one of the presenters of the first series. 'The powers thought the word "money" was a little vulgar, they thought "The Industry Programme" might be more suitable.' The forces of progress prevailed, however. *The Money Programme* was born, and was still going strong a quarter of a century later.

'What became very apparent to the journalists working on the programme,' remembers Watson, 'was that the BBC had no contacts in the City at all, except through the Board of Governors. We were starting from scratch. We also discovered that the business world was just as frightened and suspicious of broadcasters as the other way round. Business had no prestige socially at that time and it didn't welcome what it perceived as patronising attention from the broadcasters.'

THE BEGINNING OF TELEVISION TRAINING

To help get over this barrier the BBC started to offer training courses to business people hoping to appear on television, and told the *Money Programme* journalists that they would act as trainers.

'I was told to turn up at Kensington House [a BBC office in west London] to hold a course,' says Watson, 'and on the first morning was confronted by twenty Swiss bankers.'

Nothing much changed for the next decade, except that training in television techniques became a more common requirement for top management. But in the Seventies the video cassette burst onto the market. Suddenly the ephemeral things called programmes could be captured, packaged and sold as products. They could be shown to target audiences at times which were the choosing of the viewer not the broadcaster. Suddenly it became possible to think of television as a communication medium for the business world, not just an entertainment medium.

THE ARRIVAL OF THE VCR

As long as television meant just broadcasting it was a question of the outside world looking at the world of business. With the advent of the video cassette, business could itself begin to talk on television, and could begin to target specific audiences.

Initially the European manufacturers of video cassette recorders saw their market as the world of education. Their machines were expensive and their largest potential customers looked as if they would be ministries of education. But in the mid Seventies the Japanese manufacturers appeared in the market and the VHS video format became standard in the UK, knocking out non-compatible, rival systems. Prices dropped almost overnight from around £2,000 for a machine to £300. Within the next five years virtually any video would fit into virtually any video cassette recorder and it became apparent to the manufacturers that they had a major new medium of communication on their hands.

THE SPAWNING OF PRODUCTION HOUSES

In the late Seventies Michael Rodd was working for the BBC in the science and features department, most notably on programmes like

Horizon and *Tomorrow's World*. He, and many others like him, were coming to the conclusion that there was a need to create a wider understanding of the processes of commerce and industry, and he didn't believe that the BBC was doing that.

'Eighty per cent of the products which we were covering on *Tomorrow's World* were sinking without a trace,' he explains. 'We believed we were a serious science programme but in effect we were producing entertainment based on the potting-shed inventor syndrome. Few of the products we were talking about were meeting genuine needs, and the whole concept of marketing was completely alien to most of the people making them. Some of us believed that we would be serving the viewers better if we slanted the programme in a more marketing-orientated way, using the power of the screen to put across positive business stories to as wide an audience as possible.'

The BBC, not surprisingly, did not want to mess with a winning formula and Rodd left to set up Blackrod, which since 1980 has been one of the major players in the world of business and corporate television. The philosophy behind such companies is to use television as one communications medium which industry and commerce can use to put across its messages to chosen audiences.

The business world in general was slow to see the potential which television was offering. Everyone thought of television as an entertainment medium, despite the launch of the Open University and other projects. It was also still élitist and expensive. The nearest that most business people got to programme making was the creation of their commercials, which were enormously expensive and far removed from the everyday world of business communication.

THE CORPORATE FILM UNITS

Some larger companies had 'corporate film departments', which were extensions of public affairs but were not tied in to any sort of marketing or human resources strategy.

'Business television was largely about making self-indulgent movies,' says Ray Townsend, chairman and managing director of World Wide Pictures, a production company which has been in the business since 1942. 'No one gave any thought to how the end products could be used. I remember Rothmans once asking for a film on *Great Bridges of the World*. It was a beautiful film, but had nothing to do with their business needs. Business television only comes of age when people learn how to harness its enormous powers properly.'

One of the most extensive corporate film units was within Shell. It was set up before the war to make 16mm films of general informational value for employees and the educational establishment.

'We had units dotted around the world in places like Singapore, Lagos and Venezuela,' says Maxwell Michie, head of Shell's Film and Video Unit. 'I like to think of them as the last vestige of the Victorian amateur gentlemen filmmakers. They worked to very high standards, but there were no commercial pressures on them.

'We had very strong ties with educational authorities around the world and the films suited that purpose perfectly. Some of the films were explaining the nuts and bolts of how our industry worked, others were made on a grander scale, looking at global problems. We made a film *The Rival World*, which was about locusts and the threat they posed to man, we did another called *Food and Famine*, and another called *Fate of the Forests* on the dangers of deforestation. These were all environmental subjects but they were being made years before the pressure groups came into being.

'We weren't using the films to any sort of promotional ends, they were simply designed to foster good relationships, and we still make films like that, alongside the more commercial material.'

THE INDUSTRY MATURES

'On the whole companies are now very hard-nosed about the business,' says Townsend, 'and only make programmes which are an investment for the company. They have to know that a programme is going to have an effect on the people who see it, there is no point in doing it for its own sake.'

Slowly, however, managers began to see ways in which they could use television to talk to different audiences such as employees and customers. Video production houses began to spring up to meet the new demand, creating corporate videos which were the television equivalent of printed brochures and annual reports. The very largest companies actually installed their own studios and film units to make their own programmes, at enormous expense.

Some very successful video training products came on the market, particularly the John Cleese management training films from Video Arts.

THE NEED FOR PERFORMERS

The gradual expansion of the industry created a need for business people who knew how to appear on the screens and communicate convincingly, and the trainers grew and expanded to fill the need. Finally, in the late Seventies the corporate television industry burgeoned and in doing so it behaved like most immature sectors.

There was a lot of learning from experience, and there was a lot of money wasted on ill-conceived projects and over ambitious promises. Production companies came and went. At times it looked as if a new licence to print money had been issued. At other times it looked as if a scam was under way which was soon going to run out of steam.

THE ARRIVAL OF SATELLITE

With the Eighties, however, came several new forces. There was the rise of the satellites into space, making it possible for corporations to create their own dedicated networks, (known in the trade as 'business television', but in this book referred to as 'private networks' or 'corporate networks', to avoid confusion).

Cable networking was spreading at the same time, as well as developments within the computing and telecommunications industries which linked television pictures into their own product developments, resulting in futuristic concepts like video conferencing and personal computers which can carry moving pictures as well as data on their screens.

AND NOW FOR THE REVOLUTION

At the beginning of the Nineties we are only just at the start of the television revolution. Just as mainframe computers were with us for several decades before the personal computer revolution became possible, so broadcast television has formed the basis from which everything else is possible.

At the end of the Eighties the International Visual Communications Association (IVCA) was formed to represent the corporate and business television industry, with a brief to encourage companies to use visual communications as part of their marketing strategy, rather than just tacking it on the end if the budget allows. Their contention is that all business television projects should be planned like commercials, which are a major part of any product strategy, integral and planned from the start.

In 1990 the IVCA undertook a survey of their industry in the UK in conjuction with the CBI. It found that the industry was worth a total of £377m, comprised of £260m for commissioned film and video products, £85m on in-house production, £24m on purchasing or renting generic training products and £8m on interactive video production. This represented a growth of 25 per cent on the previous year.

During the year the IVCA estimated that around 10,000 programmes were commissioned by British industry from production houses, and a further 6,000 were produced by in-house operations.

Respondents to the survey were asked to indicate in what area of application their programmes were made, and they fell into the following categories:

	per cent
Training	28
Employee communications	19
Sales promotion	18
Public Relations	12
Conference/Live events	9
Public service/Education	6
Shareholder/Financial reports	1
Other	7

'Other' commonly included functions like recruitment, research and development presentation, confidential research, membership and promotion of services. There is also corporate health and safety, medical, public welfare and safety, video publishing and interactive videos to add to the list.

The percentage of respondents commissioning programmes in each area fell into the following categories:

	per cent
Training	40
Employee communications	36
Sales promotion	34
Public relations	31
Shareholder/Financial	6
Public Sector/Education	19
Other	11

All the research pointed to the fact that the two areas where the greatest concentration of effort and money goes are training and corporate image programmes.

The average annual expenditure on commissioning programmes within the survey samples was £86,000.

Expenditure	*per cent*
Up to £25,000	26
£25,000 – £100,000	53
£100,000 – £300,000	15
£300,000 – £500,000	2
£500,000 – £1m	3
More than £1m	1

The average programme budget for the same sample was £24,700.

Budget range	*per cent*
Up to £10,000	20
£10,000 – £15,000	17
£15,000 – £20,000	14
£20,000 – £25,000	8
£25,000 – £30,000	15
£30,000 – £40,000	11
£40,000 – £50,000	8
£50,000 upwards	7

ESCAPING FROM THE MASS MARKET

So what has happened in the world of broadcast television while all this has been going on?

Deregulation is what has been happening, and with it has come opportunities for creating specialist television niches aimed at valuable target markets, one of which is the business community.

As broadcast television has become the most powerful and pervasive medium in the world, the business community has equated – quite rightly in many cases – mass market with down market. Most of the programming seemed to be aimed at housewives, teenagers, children and sectors of the public other than themselves, people with seemingly large amounts of time to spare and a need for escapism and entertainment.

Because of their lifestyles, most of the upper echelons of the business world don't have time to watch the traditional programmes anyway, certainly not on a regular basis or in their homes. They come home late, they leave early, they spend large amounts of time on aeroplanes and in hotel rooms.

With government deregulation allowing independent broadcasters and programme makers to compete with the traditional monopolies of the BBC and ITV networks, and the launching of satellites and cable networks, that situation changes. Suddenly there are different ways and means of reaching business people, which means new possibilities for the advertisers and marketers, and new opportunities for the media owners. The television set is no longer static in the corner of the living room, or possibly the bedroom, at home. Now sets are available in offices, in hotel rooms, in airport lounges and even in the back of limousines and inside aircraft. Video cassettes can be carried around just like magazines and newspapers, and the English language has become the business language of the world.

Although newspapers like the *Sun* can win the biggest readerships because of the mass-market appeal of their editorial content, the *Financial Times* remains the most profitable newspaper because of the value of the target audience that it reaches. There may not be as many *FT* readers as *Sun* readers, but they are people of power and influence and are consequently the sort of people that everyone wants to talk to. Because they are hard to reach it makes the rewards for the media which can catch their attention all the greater. The same laws of audience segmentation now apply to the world of television.

TARGETING THE BUSINESS MARKET

All the media owners now know that if they want to grow to be giant players in the future they must have as many fingers in as many media pies as possible, and television is the juiciest potential pie of all. There are still expensive mistakes being made by the pioneers, but they are mistakes worth making, because the potential rewards in the future are limitless.

The *Financial Times* is a good example of the way the business community is going to be served by broadcast television in the future. In 1989 Financial Times Television was established by the Pearson Group to produce current affairs and business programmes for an international audience. The viewers they were aiming at were leading business men and women across the world, particularly Europe – the same people in fact who read the paper.

If there is one thing that the satellite companies lacked at the beginning it was credibility amongst the business classes. If there is one thing the *Financial Times* can provide it is credibility. There seemed to be a mutual need to be satisfied. Programmes with the *FT* stamp would attract the right types of viewers to the burgeoning new television station.

They started by producing *Financial Times Business Weekly*, providing a weekly half hour programme in English, covering main business news stories, backed up by regular features. It has the magazine format which had been largely invented by the BBC's *Money Programme* and had been developed by others like Channel Four's *Business Daily* programme.

Broadcast on Superchannel on Wednesday evenings, with a repeat and up-dated edition on Sundays (the programme also appears in Japan on Tokyo Television and in America on the Financial News Network), it claims to concentrate on interviewing chief executive officers and government ministers rather than the analysts from stockbroking companies and merchant banks who had become the staple diet of other business programming.

Because of the cachet of the *FT* newspaper, these leading figures were more willing to co-operate than they might otherwise have been. In the first year the programme did exclusive interviews with the chairmen of Philips, Olivetti, Porsche and KLM, and many leading politicians.

In 1990 the *FT* moved a step further going into a co-production agreement with Cable News Network (CNN) to produce a daily programme *World Business Tonight.* It is made at the *FT*'s studios in London, and is part of CNN's moves to counter criticism that it is an American station which happens to broadcast around the world, and to create a genuinely international product. It consists of two half hour nightly business programmes which are fitted into CNN's 24-hour news service.

As media opportunities for advertisers, these programmes are catering for the same audience as magazines like *Time*, *Business Week* and *Newsweek.* They reach a very elusive and expensive target market and are useful to advertisers with global messages like the major airlines, car manufacturers, oil companies and parcel delivery services.

'Business people are very hard to reach,' explains Roy Duke, sales director of New Media Sales, a company which has spun off from Superchannel to exploit the burgeoning satellite market. 'Traditionally television advertisers could only talk to them through the breaks in major sports programmes and *News at Ten*. Sometimes it was possible to reach them on Sunday afternoons. But there was no consistent pattern to their movements.

'Most CNN viewing is done in hotel rooms, which is where modern business people spend a great deal of their time. They are usually in their rooms between six and eight in the evenings, and then after dinner from 10.30 or 11.00. During those periods they are likely to be looking for international news and relevant programming which they are unlikely to get from local stations broadcasting in local languages.

'We have done some viewing studies and we find that on average these people watch half an hour in the morning and another half an hour in the evening. Because of the way the programmes are scheduled in half hour cycles, they can receive a complete, up-to-date round-up of world events.

'Major news stories, like wars or stockmarket crashes, create a tremendous market need for continuous news coverage. World leaders do not want to have to wait until six or nine o'clock before they can be updated on events, they want to be able to find out at any time. If world leaders are using a channel like CNN, you can be sure that the business community is there as well.'

As a further development the *FT* is also producing *Business Report* for Superchannel, which is a five minute round-up of the day's hard business news, transmitted four times every weekday morning between seven and eight o'clock. Its purpose is to provide the European executive with an update on the major business, financial and economic stories of the hour, coupled with an overnight market report from Tokyo, Wall Street, the European bourses and international capital markets, plus a short, topical interview.

The *FT* isn't the first to see the potential in this market. An operation called European Business Channel was launched in Zurich to produce business programmes for the whole of Europe, but went broke in the attempt. The chances are that others will founder before the formula is completely mastered, although it is doubtful if the *FT* will prove to be a casualty, given its track record and resources.

CONTROL OF TELEVISION CHANGES HANDS

'Until now the cost of televisual communication has been high,' says Patrick Scott, managing director of BSB Datavision, the telecommunications subsidiary of British Satellite Broadcasting, which was merged into Sky Television during 1990, 'with control resting in a few hands. The advent of private satellite communications will bring a change to the way business systems are designed, and a new opportunity to seek competitive advantage.

'The edge in communications now equates to competitive advantage, or competitive disadvantage for those who don't harness change. The limits of advantage lie only in the imagination of the users, not the technology.

'Communications has come out of the dark ages and into the space age. Satellite-based private communications services could have as great an impact on business as the introduction of the fax machine or cellular telephones.'

This is where we have got to at the time of writing this book. From here on, things are going to continue changing at an ever faster rate as dishes spread across the roofs, cables snake their way underground and more and more people get access to the airwaves through their telephone and computer hardware. It is all a long way from the pioneers at the BBC half a century ago.

Chapter 2

Who Does It? How To Choose a Production House

Any company that does not have an in-house facility, which will always mean most companies, is going to need independent production houses to make their programmes. Even companies that do have their own studios still use outside specialists for a percentage of their work.

It has to be said an awful lot of bad business television gets made, for a number of reasons. Firstly everyone thinks they can do it, because everyone has been watching television all their lives – just as everyone thinks they can run a restaurant or a pub.

Secondly, because it is an immature industry there are virtually no barriers to entry. You do not need a great deal of capital because everything is available for hire, and you don't need all that much know-how in order to win the work; all you need is the self-confidence to go after it. As a result many companies come into the business and then go again because they are unable to win any repeat work from their clients.

This means that the whole field is enormously price competitive, which makes it hard for people to run profitable, growing businesses.

Because business television has not, until recently, been taken very seriously, few companies have people in-house with enough expertise to be able to buy services wisely. In the more mature – or serious – businesses like computing or advertising, the importance of having proper business disciplines is well recognised. Regarding the use of business television, that has still to happen in many client companies.

LACK OF CONTINUITY

As a result, most corporate television is still commissioned on a very short-term, project-by-project basis, which means that there is little continuity because different people do the commissioning each time and different production companies win the business each time.

There are hundreds of companies claiming to specialise in corporate television and videos, ranging from one-man bands, to large, sophisticated operations. Some of them do nothing else, others are just tacking corporate work onto what they really want to do, in order to bring in some extra revenue.

For someone who has no experience of the marketplace it all seems confused and confusing. The production companies are having to sell hard in order to stay afloat in a competitive business, and most of them will make claims for being more than just technicians and makers of programmes. They will tell you they are 'marketing people', or 'communications people', and they will assure you that they will only recommend 'solutions' which are right for you – in other words they promise not to make an expensive programme which will end up being a white elephant.

In some cases their claims are true. There are companies which are well set up to advise their clients on general communications needs, and it isn't always the biggest companies with the glossiest sales literature which house the most creative and competent people. There are also a number of smaller production houses which are extremely creative and adaptable. In other cases they are just a bunch of competent technicians who will need all the creative 'business' input to come from the client's side.

LARGE OR SMALL?

The temptation for any corporate buyer of services like these is to opt for a company which looks permanent and safe; something, in other words, which looks a bit like themselves. Contractors who have large offices and comforting sales literature, companies that can boast things like sales forces (or 'new business managers') and in-house production managers, make them feel more comfortable.

In most cases they will be okay, they will have made a safe choice by going for the biggest. As they say in the computing industry, 'no one ever got fired for buying IBM'. They may also, however, be making an expensive choice and they may be missing opportunities for working with more flexible and adaptable operators who could become an almost integral part of their own marketing team.

A FREELANCE TALENT POOL

Even the big production houses, however, survive to a large extent on freelance talent. The producers may be in-house, but the chances are that the directors and writers will be pulled in from outside, and this is probably the best system for keeping levels of creativity high.

'I have five producers here,' says Ray Townsend of World Wide Pictures, 'but I don't want to be tied down to using a director or a writer on a project just because they are on the payroll and haven't got anything else to do that week. I want to be able to hire the best people for each job. If a client needs a comedy film, the producers know who the half dozen best comedy writers in the country are at the moment, and will probably have some sort of relationship with half of them. They can then find the best person and present them to the client.

'In some cases they can be rather unusual characters, and it is the producer's job to act as a buffer between the client and the creative people.'

In the final analysis you are hiring individuals. Once you have found someone you trust who will take the burden of making a programme off your hands and then come back with something which (*a*) complies with your brief and (*b*) does the job that you hoped it would, you won't care whether they belong to a company with a suite of offices in Park Lane or if they work from their spare bedroom at home.

Once you have found the key person that you want to work with, someone who understands your needs, you can then leave it up to them to pull together a team of technicians who can create the product that you want.

HOW DO YOU FIND THE RIGHT PEOPLE?

So how do you set about finding someone with whom you are going to be able to form a long-term working relationship?

In their 1990 survey the IVCA asked companies how they came to hear about the film or video production company that they last used. Thirty-four per cent said they had used them before, (against 43 per cent the year before), 26 per cent said they had been recommended by a colleague or acquaintance who had used them before, 19 per cent said 'word of mouth', 6 per cent had responded to a direct sales call, 3 per cent to a direct mail shot, 2 per cent had gone through their public relations company, and another 2 per cent through an advertising agency. Eight per cent had used other methods such as trade shows, sales agencies, professional bodies like the IVCA, and editorial and advertising in magazines.

Asked to list their priorities when choosing a production company, respondents put 'an understanding of business aims' at the top of their list, followed by 'creativity', 'cost/value for money', 'planning ability', 'personnel', 'track record', 'reputation', 'experience in the client's industry' and 'awards won'.

Eighty-four per cent of the sample commissioned its programmes directly through the independent production houses, while small percentages went through their advertising agencies and public relations consultancies.

HOW MANY TO APPROACH

'A lot of clients approach as many as six production houses for pitches,' says Roddy Gye, managing director of GHA, an independent production house, 'which is too many. Three or four is a better number, and is fairer to the production houses.

'The prospective client should give us each a written brief explaining their problems and their audience,' says Gye, 'and we can then go back with a creative response. We have to be able to understand their objectives so that we can do some sort of analysis of the influences at work and see which way we should go. We can then suggest a broad theme.

'We always prefer to pitch face-to-face, and although we accept that they need to see samples of our work it would be a mistake if they were to make a judgement just from a showreel. The temptation is for clients to treat showreels as catalogues of techniques.'

Many of the production companies will be able to send showreels in response to an enquiry, but they should be treated only as demonstrations of a level of competence within a company. A client who picks a video style from a reel and says 'I want one like that' is going to end up with something which looks perfectly professional, but which won't necessarily be the best answer to their problem.

Some client companies will ask production houses to do a creative treatment on spec, and they will then choose the company which gives them what they believe to be the 'most imaginative' ideas. That is almost certainly a mistake, since a skill for writing imaginative treatments on subjects of which you have only a superficial understanding is a long way from writing scripts that will answer the problems and needs of that particular situation and actually 'work' for the client.

EVERYONE'S AN EXPERT

It is also impossible for someone who is not experienced in television to be able to look at a script and be able to envisage how it will work on the screen. But because everyone watches so many broadcast programmes, they all think they know what is needed.

'An awful lot of production houses are led by creative people,' says Martin Pentony-Woolwich, a director of Spectrum Communications. 'They start with producers and directors and then they bolt on all the other services. They are not often marketing people. We believe that it is better to start with a marketeer who can act as a creative filter between the client and the others. There are a lot of gorgeous-looking videos being made which do not do any useful marketing job for the client who paid for them.

'Business television is not being made for light entertainment, it is there to do a job. In order to do that job, however, it will need to be creative and stimulating, saying what needs to be said in an interesting and informative manner.

'When buying television and video services clients should always look for experience, and ask to see samples of the production company's work. Clients should ensure that they meet the actual people who made the programmes they are being shown – sometimes a busy production house will hire a freelancer to make a programme which it then shows around as its own work.

'When talking to them find out what their problems on the production were, the way the costs worked out, and ask what they enjoyed most about doing the job. Generally get to know how they deal with problems and come up with solutions.

'In the end the client is judging the work, and should take no notice of any mitigating circumstances. It is not relevant that "the light was fading that day" or "the client changed his mind half way through shooting", there are always going to be mitigating circumstances and what matters is whether the production team were able to produce work to a decent standard in spite of all the problems they had to face.'

Roddy Gye of GHA says, 'Most clients are not experienced at making television programmes, and so they conform to stereotypes, wanting to remain within the conventions of what they have seen before. Although that is an understandable attitude, it stifles creativity.

'Every programme needs to be individually tailored to solve specific problems and to do specific jobs. Provided you pick a company which is technically capable of making a programme to the required quality standards, the only other criterion which should worry you is, do they understand the needs and style of your company? You will only be able to tell that from meeting the people who will actually be making the programme. A showreel or sales person will only be able to convince you of their technical capabilities.

'The first time that a production house works with a client, they need to go in and fully familiarise themselves with the client's operation, just like an advertising agency or a public relations consultancy. If they then manage to build a relationship they can bypass all that preparation work on later projects because they will already have a basic understanding of the client's needs and the ways in which the business works.'

Some client companies do offer to pay rejection fees, which could range from £50 to £3,000, to production companies when they reject invited tenders. It is a practice which the industry, obviously, wants to encourage, since making pitches can be an expensive and time-consuming business, particularly for the smaller operators, but it is still a long way from becoming standard.

NO SUCH THINGS AS BARGAINS

'Clients must accept that there are no bargains available in this industry,' says Ray Townsend of World Wide Pictures. 'Good people cost money, and so clients will always end up getting what they have paid for.

'Ideally they should tell us how much they have to spend, and then we can suggest the best way to use the money to solve their particular problems. If they don't name a figure at the beginning we have nothing solid on which to base our suggestions.

'Do we suggest a programme which will cost £50,000 and will be highly effective at solving all their problems? Or do we suggest one at £10,000 which will solve one aspect of the problem? Or do we suggest that television is completely the wrong way for them to go because their budget could be more effectively spent on print.

'Once they have a shortlist of possible production companies they should ask for some names and addresses of referees. At least one should be of a client the production company has worked with before. Another should be the production company's bank, and another their auditors.

'It is vital that the company is solvent if the client is going to do business with them. If they go into liquidation towards the end of the project the client could end up losing a lot of money, and possibly their programme as well.

'Nobody would hire a trucking firm without looking pretty deeply into the way they do business. Why hire a production company on the strength of their sales talk? Television suppliers should be treated like any other potential supplier. The reason this doesn't happen is because the client companies undervalue the role of television.'

Once the client has decided to go ahead with a production house, they then have to decide on what sort of legal agreement to come to. Only 44 per cent of programmes produced are made with any form of agreement or contract and 43 per cent are produced simply on a letter of agreement. The IVCA is not happy with that and would like everyone to use standard agreements for the protection of both parties. But although it is easy to see the benefits of such practices, it is sometimes easier to build up relationships of mutual trust without complex legal agreements.

HOW TO COMMISSION

Aware that clients who have not been involved with television before do not know where to start, some of the larger production houses hold educational seminars to explain how to commission. At Blackrod they call them their Video Maze.

'We start by explaining that there are around 1,000 production companies in the UK,' says Graham Higginson, Blackrod's director of business development. 'The majority turn over less than £500,000 a year, around a hundred of them turn over less than £1.5m and about twenty-five turn over more.

'A small company may have no financial security and may be running to stay ahead of the bank manager. It will probably be a producer-led hotshop putting together a freelance team whenever there is the chance of a job. Some people are more comfortable with that, believing that they are getting a better service and that they will be able to impose a greater influence over the production than they would with a larger company.

'The larger companies are going to be more establishment based, with on-staff production management, proper accounting and security systems. The client may have to pay a little bit more, but they will get a lot more, and because of their size the larger production houses can negotiate better rates with suppliers because of the volume they are buying.

'So the clients must start by deciding whether they like the people who are pitching and whether they could work with them. They need to be sure that the team understands their business. There is a lot of arrogance in the television industry, amongst producers particularly, and clients should avoid anyone who takes a high-handed attitude.

'Next they need to ask themselves about the team's creativity. Will they, in other words, be able to reach the goals being set them?

'Then there is the production management side. Can they provide the organisation and management facilities necessary? Will they be able to bring the project in on time and on budget? A small company, for instance, may not be equipped to handle more than one project at a time. If something else comes up while filming is under way the main players may have to go off to deal with it. An awful lot of productions are delivered late because the producer had to go out selling other projects.

'Financial security is also important. The worst thing that could happen is that the production company goes bust halfway through the project, having taken a large part of the money up front.'

Blackrod's Video Maze also lays out a definitive commissioning and execution process which, with variations to suit every situation, provides a typical scenario which any client working with an established production house will meet. It starts with the project instigation, at which the client defines the objectives of the programme under discussion, and the production house finds out exactly why the client wants to do it. They then identify the target audiences and develop a strategy on how to get the product to them.

Any video or television project must relate to the company's other marketing or training activities. If it is made in isolation it will be useless. So the production company needs to know what else is going on within the client organisation and how the programme will fit into it. They need to look for ways to get added value from the programme.

There are then the practicalities to be considered. If, for instance, the company wants pictures of its factory in Sweden, it is no good going out to film in the middle of winter. The timescales and the budgets need to be discussed to work out what is practical and what is ideal.

At this stage Blackrod will produce a creative treatment, which will lay out the basic format of the programme – is it going to be drama or documentary, for instance – with an outline of the action and some test dialogue. If the client wants it done on storyboards then they will charge for the service. A schedule will be prepared to demonstrate how long will be needed for shooting and editing.

There might still be two or three other companies pitching. 'If there are more than that involved in the pitch we might decline to join in,' says Higginson, 'since the chances of winning are diminished, unless we feel we have a particular angle which we are confident will win the day for us. In those cases we will be happy to send our credentials' documents and invite the prospective client to come back to us if they want to shortlist companies.'

If the company wins the job it will want 50 per cent of the fee at this stage, and then it will produce a shooting script, with the writer, producer and director working on the project on a daily rate. Actors will be cast, designers briefed, crews booked, graphics planned, existing materials evaluated and locations and contributors identified.

'We will do all the research for the client,' explains Higginson, 'finding people with something to say, and talking to the people they suggest, to see if they will be suitable. Sometimes there might be people they haven't even thought of talking to, like satisfied customers.

'We also ensure that everything we do receives clearance from the right people within the client company. We will never allow a crew out until we have all the necessary authorities and permissions, otherwise you can end up wasting a lot of time and money.'

It is also important to keep control of where copies of the finished programmes go. An oil company, for instance, which suffers an oil spillage that gets into the news, will find that the news editors will dig up filmed material of the company from any source they can.

When all the planning is completed it is time to start on the actual production, and at that stage Blackrod will ask for another 35 per cent of the agreed fee. They will then shoot the film, create the graphics, process the film rushes and arrange for the clients to view them. After the post-production work, and with the client's approval of the finished product, the final 15 per cent is payable.

DISTRIBUTING THE PROGRAMMES

Just as people are becoming used to the idea of dealing with television production teams, talking to cameras and working out budgets, a new strain of techno-fear has arisen to take its place.

The sudden availability of new distribution methods – most notably satellites – has meant a bewildering new set of international regulations and technical jargon to be confronted. It often seems easier to ignore the whole thing and continue doing things as you always have, than to risk things going wrong. Again it is every client's dream to find someone to whom they can hand over the whole package.

Creative Technology Ltd. is a company which supplies the hardware and the people to use it, generally working through the production companies which stage major events like satellite transmissions for corporate conferences, product launches and privatisation shindigs.

'We can supply them with everything from a twenty-inch monitor to a twenty-foot daylight screen,' explains Graham Andrews, Creative Technology's managing director. 'We can supply outside broadcast vans or video graphic equipment for presentation aids. But what is more important for many clients, we can organise all the ups and downs to the satellites.

'The clients need people who can originate the pictures, which we do; organise the transmission, which is basically a broking job; and then organise the display of the pictures at the other end.

'We are very conscious that clients are nervous about getting into deep water with things like international conferences, and will often revert to flying everyone around the world, because they are comfortable doing that, when they could be saving themselves hundreds of thousands of pounds by using telecommunications and television technology.'

This unease is partly the fault of the business television companies, thinks Bob Clarke, chairman of Visage, the communications company responsible for creating the first major private network in Britain, for BMW. 'There is enormous confusion in people's minds, and there are probably only about 300 people in Britain who really understand what business television is.

'The clients have been further confused with talk about satellites, up-links and down-links, when they don't need to know anything about the technicalities. All they need to know is how they will be able to use television to improve their corporate communications, and then leave it to people like us to make it happen for them.'

SCRIPTWRITING

Sometimes clients make the mistake of believing that they can write their own scripts, or that they can transfer words which have been written for a corporate brochure or a speech onto the screen. Making words sound natural when spoken to a camera is very different from the literary skills needed to bring prose alive on the page. It is closer to speechwriting, but even that is not quite the same thing.

Sometimes scripts will be written by producers or directors, sometimes by full-time writers. All of them are professionals in the art of making people look and sound good when they appear on television. They know how to make dialogue sound natural, and they know how to simplify messages down to their purest essence.

The pictures should provide the illustrations, so the words do not have to embroider further with descriptive adjectives and adverbs. Dialogue needs to be natural, voice-overs need to be spare and factual, informative and clear.

Too many images and messages will confuse the viewers and they will be unable to remember the key messages at the end. The eye works faster than the ear, pictures really are worth a thousand words; the words merely have to provide the abstract information which explains what the viewer is seeing, if the picture is not self-explanatory.

In some cases scripts have to be funny, in which case they need professional comedy writers. Sometimes they must be written in the style of a particular type of programme, which requires a specific expertise.

Chapter 3
How Much Does It Cost?

PRODUCTION VALUES

'When you hire someone to make a video of your wedding,' says Roddy Gye of GHA, 'you probably have to pay them £1,000 or so. For their fee they will make sure that everyone gets on the film, that there is a picture of the cake, and the speeches are well recorded. It will be a nice memento of the day – which is exactly what you wanted.

'When Prince Charles married Lady Diana Spencer the television companies spent millions of pounds creating a programme which would be beamed out around the world to the biggest audience ever. Everyone who was watching was expecting to see a programme of the same standard as their favourite broadcast programmes. They expected to see streets lined with people, majestic aerial views of the cathedral, and close-ups of the people involved. They expected the cameras to stay with the couple every inch of the way, and they expected to hear the familiar voices of famous narrators and commentators. They too got exactly what they wanted.

'What it comes down to is "production values". You have to decide what you want your programme to look like. There is a general belief that you can cost-out television programmes by the minute, but costs bear no relation to running time, they are simply governed by quality.'

Although the example is somewhat extreme, it illustrates the point. If you are making a corporate video, for instance, you could make it to the standards of a local news broadcast, with one camera and some simple, quickly set-up, shots. Alternatively you could produce something of the standard of some of the blue-chip television commercials, where it might cost £100,000 to create thirty seconds of breathtakingly beautiful television.

WHAT SORT OF PROGRAMME DO YOU NEED?

First, therefore, you have to decide what sort of programme you need, and consequently what sort of technical standards it must meet.

Secondly, you need to know how much you can afford to spend on the project. If the two factors are wildly incompatible, then perhaps television isn't the answer to the communications problem you are dealing with.

There are several elements which can push the costs of television up out all proportion to the benefits. Using a star name from the world of broadcast television will always mean paying an exorbitant premium, whereas it is always possible to find a less well-known professional for a fraction of the cost who will do the same job, but who will not provide the cachet.

Straight narrative or studio discussion will always be cheaper than a drama which involves actors, costumes, and all the attendant expenses. Shooting in more than one location adds enormously to costs, since you are having to pay for extra travelling and setting-up time for everyone on the team.

High-technology animation and graphics, with all their visual tricks, are very tempting, but can also be very expensive. It is possible to buy certain packages 'off the shelf' at considerably lower prices than having something tailor-made by an expensive facilities house.

There are many other ways in which you can keep costs down during the editing and production process, which a producer or director will be able to advise you on at the beginning. You need, however, to be clear in your brief about how much you want to spend and which aspects of the project are important.

THE COST OF PRIVATE NETWORKS

The costs involved in putting together private networks for broadcasting to selected closed user groups are still very hard to pin down. What is safe to say is that it is almost certainly not as expensive as you might think.

The pioneering companies involved in the sector, like British Aerospace, Visage and Creative Technology, are all keen to prove to the business community in general just what a useful communications tool a private network can be. None of them, therefore, are looking for quick profits; all of them are looking for clients with whom they can strike up long-term, fruitful relationships, which will provide talked-about showcases for the new communications technologies.

If you are a major user of corporate videos, the private networks may provide a viable alternative method of distribution, and might even lead to cost savings in the long term.

IN-HOUSE UNITS

One way of taking control of the costs is to create your own production facilities in-house. This could mean anything from having a single person in charge of all the video products of a company, to complete studio and production facilities. Some large companies like Shell, Wellcome and Barclays have been doing this for some time to varying degrees. It isn't always as cost-effective for smaller organisations, although Unipart have found it extremely successful at their Cowley headquarters.

'Every month we make a half hour programme for our major customers called *Channel 5*,' says Unipart's public relations director Patrick Fitzgibbon, 'updating them on the products and telling them about special offers and promotions. We make another half hour programme each month for employees, called *Grapevine*.

'We used to use outside production houses, but it simply became too expensive and we decided to bring it in-house, everything from scriptwriting to final production and editing. We have everything here, including an outside broadcast van. One month, to celebrate our 21st birthday, we broadcast *Channel 5* live from the studio to eighteen different locations where we had gathered nine hundred of our customers together, and they were able to feed back into the studio live discussions down the telephone lines.'

Even with that much activity, however, the studios are not fully utilised just on Unipart's own business, and they are hired out to clients like the RAC, with the Unipart team acting exactly like a freelance production company.

The costs of setting up an actual studio are no longer that high, and are falling all the time as high technology becomes more accessible and more competitive. The problem may be obsolescence, with new gadgets coming onto the market all the time and making what went before seem redundant. If a company is just making programmes for internal consumption, that may not matter too much, since good technicians often work better with equipment they are familiar with than with something new. Alternatively, a company might decide to lease and hire hardware whenever they need it, but use in-house expertise to do the work.

Which method works out the most cost-effective depends completely on the programming needs of the company.

THE COST OF GOOD PEOPLE

What never gets any cheaper is the people, and it is the cost of hiring expensive full-time production teams which puts most companies off starting up their own in-house units.

To make it pay, you have to be able to keep them busy the whole time, and in order to do that you have to ensure that all the company's film and video work goes through that department.

Once a company reaches a certain size, that begins to become less realistic. British Telecom, for instance, once had an in-house film and video unit of eight people. The company is so huge, however, that it is quite likely that many of the people needing to have a video made did not even know that the department existed. There was also no way that a team of eight could handle all the work needed by such a large organisation. The result was that they were actually competing for the business against some very sharp private sector companies. What possible motivation could an in-house team have to win business when everyone is on a salary? It is very easy for them to become inward looking.

In a round of cost-cutting, the team was cut from eight to two (an easy cost to cut without anyone noticing) and suddenly the department did not look viable. There was a similar story within British Rail.

For an in-house production unit to work, either they need to be handling all the company's work – which will suggest the company is of a certain size – or they must have a specific task, perhaps to create all the training material or a daily video magazine. If the brief isn't tight, they will soon become a backwater.

IN-HOUSE AT WELLCOME

Wellcome, the pharmaceutical giant, has a complete production facility called Wellcome Video, based at Chesham.

Says the manager, Andrew Brown, 'We act as producers for any clients who come to us from within the company for whatever reasons. It might be from public relations, marketing, employee relations, training or for gathering news material for sending out to the broadcast companies. We work within the public relations department, although we are self-contained and entrepreneurial, with five full-time staff. Just like independent production companies, we budget and charge out all our work to the "clients" within the group.

'No one is obliged to use us if they would rather go to an independent, but hardly any of them do. Sometimes they work with people like London Scientific Films on specialist projects, for instance, and we are always happy to co-operate with things like archive material. It is important that we offer a competitive service, and do not do projects on the cheap.

'We did have a studio, but we closed it because it was under-utilised because we do so much of our work on location. We have an on-line editing suite, which produces work to broadcast quality. It is extremely convenient to have it on site, and it provides us with a high degree of confidentiality, which is important in the pharmaceutical industry. It also means we can take time tinkering with productions and dubbing them into foreign languages, which would become very expensive if we were hiring facilities on an hourly basis. In the short term it means a big capital expenditure, but in the long term it must work out cheaper.'

If they are doing a particularly high budget production, Wellcome might go to outside facilities which have better equipment, but it is seldom necessary. The same philosophy applies with the actual filming equipment. Most of it they can do themselves, but sometimes they need extra pieces of equipment. They have three in-house producers, but they hire writers, directors and camera crews as and when they need them – just like the big independent companies.

'Another advantage of an in-house production unit,' says Brown, 'is that we can keep master copies of everything we have made, which we can distribute when requested. We put out a catalogue of programmes available, and people ask us for whatever they need. It is a strong resource for a company to have.

'Our training departments don't use us all that much, as each of the main training sites has its own video facilities, putting together its own material to lower production values than ours.

'Different companies have different reasons for setting up in-house facilities. Only about half are set up for specific purposes, and are given budgets with which to achieve pre-determined goals, the rest just seem to evolve. I have come across a number of other companies which are approaching the whole area of in-house video very professionally. General Portfolio Assurance, for instance, provides a very thorough monthly video magazine to all its branches, and VAG used to do a lot of servicing material on video for its garages, although they have now gone independent.

'Some in-house departments keep going along very steadily for years, but others become entrepreneurial and start to take on other work, and eventually float off to become independent.

'In some cases they become too big and then get axed. What happens in those cases is that they enjoy some early success, and so the company concerned decides to invest more in hardware and staff to make them even better. Before long they are costing the company £1m a year, which is fine as long as the good times last. As soon as hard times come, however, that is an easy way to save a million pounds a year.

'To be successful, an in-house video facility has to work hard at keeping a high profile within the parent company, and has to be careful to remain as good, or better than, outside providers of the same services.

'The advantage which we have over outsiders, I believe, is that we don't have to keep going over the initial learning curve with each new project. We already know about the company and the products, so we can capitalise on that knowledge by moving faster in the early stages of a production. We understand the needs of the company and the marketplace.'

IN-HOUSE AT SHELL

One of the oldest and most established in-house departments is within Shell, where the unit is a major part of the public affairs department, accounting for around half its annual budget. As well as making the general information and educational films for which they are famous, the unit also provides facilities for the making of training, safety and product information material.

When managers within the company come to the unit with a request for a programme, it can sometimes be put together from archive material. If, on the other hand, it requires the origination of new material, the unit will hire freelance creative people, just as the production houses would. In fact, they are hiring exactly the same people. So the only differences are that the producers are part of the company, with all the necessary background knowledge, and many of the facilities like editing are in house.

'One of the reasons we keep it in-house,' explains Maxwell Michie, head of the video unit, 'is to create a sense of oneness in the business as a whole, a feeling of cohesion. Shell is made up of a hundred or so largely independent operating companies, and we provide a centralised service for them.

'Other advantages include keeping the profit margins inside the company, and being able to call on an enormous pool of expertise within the organisation on subjects like mining or forestry. The problem with independent production houses is that very few of them specialise in any one subject area. When you do find one that does – and we have used one which specialises in making geological films – it can be unbeatable.

'In-house facilities can work well if you have a critical mass of work, and in America the in-house production departments far outnumber the independents.

'The difficulty in a company like Shell is that the decision-making processes are so complex because of our size. There are so many people with a say about the content of any video that we make, all of whom need access to the off-line edits, that an independent producer would be driven mad. Many of the technical subjects we cover are very complicated, and we might well have two experts making an input on a subject who disagree with each other over a technical point. It is easier to sort out that type of thing internally.'

SOME IN-HOUSE EXPERTISE ESSENTIAL

It takes a big company to support operations like these, but all companies which take their communications activities seriously should have some video and television expertise in-house, even if it consists of only one person.

Any company which knows nothing about the subject is laying itself open to being taken for a ride by independent suppliers, and can easily lose control of the public relations benefits which can accrue from film and video.

'In some large companies the person who has responsibility for buying video and television services has no experience of the medium,' says Ray Townsend of World Wide Pictures. 'The result is that television is undervalued and not seen as a key ingredient of the communications mix.

'When companies employ communications directors they expect them to have press relations experience and personnel experience, but no one expects them to have worked with television. Often executives in very senior positions have no idea of the pitfalls or the advantages of the medium. Many have not grasped the full power of television and some are missing the opportunity altogether.'

At BP they have made a deliberate decision not to invest in any equipment, but they have a team of professional producers who can also write and direct material when necessary.

'I like to think we have the brains but not the machinery,' says Chris Bratt, head of audio-visual services. 'We are involved in all the other aspects of presentation, like the staging of events, and video and film make up part of that mix. We have to be able to recognise what is needed from the material that is being shot, so that we can see how to edit it down to perform different functions. I might go to Moscow, for instance, and while there I would rough-edit the material being shot to fit into one of the group video magazines as a news item on what BP is doing. At the same time the material could be used for a much more detailed story on marketing opportunities in Russia for another in-house programme.

'To be able to do that we need to understand the needs of the company. We need to be able to originate ideas and we need to be able to understand what people need when they come to us with their ideas.

'There has to be someone inside any sizeable company whose job it is to commission television and video material, and that person must know how to get the best value for money.'

COMMUNICATIONS CONSULTANCY SERVICES

When is a consultant not a consultant? The answer might be when he has some sort of vested interest in recommending one service above and beyond others.

The major advertising agencies, for instance, which orginally had a vested interest in persuading clients to spend more money on media buying, tried during the Eighties to diversify, telling clients that because they can provide services in all the marketing disciplines they, the clients, can trust them to give impartial advice on which sorts of marketing services would most suit their needs.

Not many advertising agencies, however, would recommend the use of business television – that is, according to those who work in the production houses.

The business television community believes that the advertising agencies see them as competition, that they believe they can do the job themselves, and that they undervalue the whole idea of business television, and encourage client companies to do the same. It is certainly true that the business world in Britain undervalues the use of television, but that is partly because no one is making a very good job of explaining the benefits to them.

The companies which are having the most success in converting clients to the cause are those which already have strong client relationships for other reasons – be it conference organisation or general electronic communications services – and whose opinions are listened to and trusted.

PRODUCTION HOUSES OFFER CONSULTANCY SERVICES

The larger and more thoughtful of the production houses are very keen that potential clients should view them as communications consultants who may, or may not, recommend the use of television to meet their needs. To swallow that line a client has to know the person involved, and has to have a great deal of faith in their integrity.

'As the business television industry becomes stronger and more mature,' says Michael Rodd of Blackrod, 'more of us should find the confidence to turn business away. If a client comes to us and says they want a video made, and we can see that it would not be the best thing for them to do, we will tell them so, and we will help them to find the right solutions to their needs.

'We believe that clients should have permanent relationships with their production houses, just as they do with their advertising agencies and public relations consultancies, allowing us to help them make strategic decisions and not simply being called in to make a film at the last minute.'

'But,' says World Wide's Ray Townsend, 'it is difficult for production houses to convince people that they can advise them strategically. Ours is a strange industry which has grown up slowly, and we are all very keen to have our services more highly valued in the marketplace. That means that in many cases we are not only willing to give clients advice that is contrary to our short-term best interests, we are often willing to give that advice free of charge, since we are not geared up to charge ourselves out as consultants.

'I have sat in at high level meetings within a UK clearing bank, which happens to be a client, redesigning pay and bonus structures for managers, and deciding how that restructuring should be represented.

'When a major Scottish company wanted to change their whole management pay structure, they asked us to go up and take a brief for making a video on the changes. They thought television would be the best way to tell managers about what was happening. I didn't believe television was the best way and I told them so, and together we constructed an alternative strategy. They then asked us to invoice them. I said I didn't have any structure of charges and they should pay us whatever they thought our services had been worth. They sent a far larger cheque than we would ever have thought to invoice for.

'The exercise would have been worth doing anyway because it allowed us to forge a relationship with them which proved very fruitful in the long run. Had they told us that they were not interested in our opinions and to just get on and make the programe, we would have done so, but the programme would not have been successful and they would have ended up blaming the medium of television for the failure.'

There are some highly experienced and skilled people working in the business television production world. Any client believing that they should be thinking about communicating through the screen should be talking to them now; not necessarily to commission a programme immediately, but just to get a feel for who the intelligent players in the field are, and to get some ideas of how television would and would not fit into their communications mix.

If you are planning to build a house or office block, it is always a good idea to involve the builders at an early stage, so that they can talk to the architects about the practicality of the plans which are being laid, and offer advice based on their experience. To draw up plans and then just hand them to a builder to execute will mean that any oversights or misconceptions which the architects may have had will only be discovered once the building work has started, and by then it may be too late or too expensive to rectify them.

Chapter 4

Defining Your Audience

'A lot of corporate television doesn't work because the makers don't give credit to the audience's intelligence and sophistication,' says Peter Cole, director of Melrose, one of the most successful video training companies. 'The same applies to other corporate communications as well.'

DIFFERENT PEOPLE LOOK FOR DIFFERENT THINGS

The problem with a great deal of television is that it is addressing any number of different audiences simultaneously.

When broadcast programmes go out to several million people at a time, there is going to be a wide variance in both the needs and the understanding of those people. Different people will get different things from a programme, which suits the makers fine. Some people will find *Dallas* very exciting, other will find it very funny, while others will simply watch it for a lack of alternatives.

With business television you need to be more targeted, since your prime objective is to communicate a key message accurately and consistently to a clearly defined target group. The least that should happen is that they should be clearly defined, and a maker of business programmes can do more to find out who his audience is and what they want to see than a traditional broadcast programme maker.

Many creators of programmes seem to feel that they need to get their work seen by as many people as possible, as if that in some way will make the costs look better per head. What they are doing is applying the principles of broadcast television, which is working to completely different criteria since they are merely trying to get 'bums on seats' which they can then sell to advertisers.

AN AUDIENCE OF ONE MAY BE ENOUGH

It could be that the success of a programme could be measured by how small, and therefore how accurately targeted, the audience is. When Mikhail Gorbachev came to Britain on a tour, which included a visit to the North Sea, the Shell video unit made a film specifically for him. Their target audience was one person, but the message which they wanted to get across to him about the company was of paramount importance.

The company sometimes needs to make programmes which will be used at one meeting only, but that could be a meeting with a government minister somewhere in the world where the company is hoping to develop its business interests. There can consequently be millions of pounds riding on the successful outcome of that communication.

Not many companies have opportunities to address audiences of that seniority, but the principles of targeting still remain. If, for instance, you are making a video which is to be seen by all employees of your company, that will mean everyone from the board of directors through the sales force to the factory floor. Some of those people will be over fifty, some will be under twenty, some will be fans of *Newsnight* and *Panorama*, others will prefer Australian soap operas, some will be game-show addicts, others will like classical drama.

You are never going to please all the people all the time, so it is always better to choose the narrowest possible target audience, the people you most want to reach. Then you can decide whether you will show the programme to the others as well, in the hope that they will get something from it, even if not as much as the primary target group. You could, for instance, make a serious management video about cost savings and how to adapt the company for the future, and still show it to people on the shop floor in the hope of motivating them to change some of their ways of working, provided it is not aggressively adversarial towards their points of view.

Some of them may well blow raspberries at the screen (which may be healthier than a resentful muttering at the back during a live presentation), but others will be listening and taking in the messages.

These extra audiences have to be viewed as a bonus, on top of the management audience you were originally aiming at. If you set out to reach everyone at once you run the risk of missing everyone at once.

A CLEAR MESSAGE

Start with a clear idea of the message you want to get across and the audience you want to get it to. Then go out and find out what that audience likes. Do they like watching television? What programmes do they normally enjoy? How long do they spend watching it? What annoys them about it? What would they like to see more of?

The management team dreaming up the programme may find that their original idea for a concept is completely unsuitable. They may have wanted to make the programme look like *The Money Programme* when the audience would rather watch *Blind Date*. With creative thinking it is always possible to adapt a message to different formats, if they prove to be more suitable for communication purposes.

The research needs to be done by the people who are actually writing and producing the programme. They need to get in amongst the audience and get a feel for what those people want and like. It is not something that can be guessed at in a meeting room.

IN-DEPTH AUDIENCE RESEARCH

A company called Futura, which is part of the Trilion television group, has gone deeply into the research of audiences for business television, involving psychologists and psychometric testing procedures.

'Before doing anything for a client,' says their joint managing director Rosita Roles, 'we give questionnaires to as many members of the audience as we can locate. We want to measure their communication styles and how they like to be talked to.' The aim is to get a picture of the character of the audience, and the environment which influences them both at home and at work. If they can picture individual preferences they find that trends emerge, particularly with in-house audiences where everyone belongs to the same company and is likely to adhere to certain psychological norms.

'We want to see what they consider as work and what as entertainment,' says Roles. 'We want to know what turns them on. Do they have logical minds? Do they like a lot of statistics, or do they find them a turn-off? We want to understand exactly what makes a message get through, so that we can avoid it fracturing or breaking down along the way.'

The company works on the theory that people have 'preferred communication styles' which fall roughly into three categories; auditory, visual and kinesthenic.

'People who prefer the auditory style, for instance,' explains Roles, 'tend to be slow readers, because they pronounce all the words in their heads. The visual people will scan quickly and assimilate a visual image of things. The kinesthenic group will rely on their feelings, and to feel comfortable about communication will need people around them.

'Having said that, most people in business today have been brought up in the age of television, and are therefore conditioned to it, even if it isn't their most natural form for communicating. There will, however, be more barriers with some people than others, but the programme makers can work to overcome these barriers, as long as they are aware of their existence. They can, for instance, ensure that the kinesthenic group are given some sort of interactive device with any programmes which are aimed at them.

'When all is said and done, this is all about communicating between human beings. The screen makes it possible to do it in new ways, but it can also be seen as a barrier between people, particularly if you view television as a way of disseminating information rather than communicating.

'The danger is that the managements of companies may have a completely different culture to the technicians who are installing the television systems and networks. This can mean that the technology will outpace the human acceptance rate, which will, at the worst, create new barriers to communication and, at the best, create apathy.'

The questionnaires which Futura sends out are analysed by the psychologists and the results are worked up into a creative proposal which is designed to give the audience what it wants, or what the psychologists believe that it wants.

The results will influence everything about the production from the choice of background music to the sort of language used and even the typeface on any written material.

'What we are doing,' says Roles, 'is analysing the corporate psyche, finding out the mood and the morale with constant monitoring, and as a result we know how to structure and adapt communications programmes.

'That way we can be sure that the people putting the material onto the television screens have the same goals as those who are taking something out of the communication. Because you are always dealing with people, you need to understand their wants, fears and needs in order to be effective.

'Business television also has the disadvantage of dealing with matters which are familiar to the viewers. If you are going to show them a programme about a conference that a lot of them went to, for instance, you are obviously going to edit it to bring out the highlights and make it interesting to watch. Anyone who was at the live event will see that this is a distortion of what actually happened. They will know that it all seemed much more untidy and boring. Unless that is handled with care, the audience will start to disbelieve everything that they see on the screen.

'We are entering a new age of communication, and companies that do not research the subject thoroughly before investing may well be buying a lot of trouble, just like the companies who rushed into the computer market before they had thought how they were going to use all the hardware they were buying.'

While it is not always possible to climb this far inside the heads of the audiences, the concept – so well known in the world of marketing – of finding out what people like, want and need and then giving it to them, rather than deciding what sort of programme you want to make and then trying to persuade people to like it, is unarguably sound.

BLANKETING THE MARKET

In some cases, of course, it is simply a case of getting a message to everyone in an organisation or a target group very quickly and very effectively.

When British Telecom wanted to tell everyone what to do about their maturing shares in the company they simply sent out 10,000 videos to everyone in their workforce down to Level 3, instructing them that all their staff were to see it.

When the listeria and food poisoning scares were hitting the food trade, Marks and Spencer made a film *Too Fussy for Words*, reiterating all the safety elements which the company is famous for, and sent it to all the opinion formers they could find, from the government to the press. They also played it to customers at charge card evenings in order to reinforce the feeling of confidence in the company.

MEASURING RESPONSE

The research doesn't end with the rolling of the cameras. Once a programme has been made, it is then time to find out what people thought of it. That way a company can ensure that they don't repeat mistakes the next time.

'We put a 25 per cent mark-up onto all the services which we buy in,' says Michael Rodd of Blackrod, 'and when clients ask us to justify it, we tell them that it is partly to fund research into how our programmes are working.

'In the past corporate television and video have been seen as a last-minute panic buy. Companies will always involve their accountants from the beginning when they are making financial plans, and their advertising agents when they are planning their marketing, but for some reason they only ever call in their television production companies at the last minute and ask them to provide some gimmicky solution to a communications problem.

'We should be part of their strategic planning process, and that means we need to know what is working and what isn't, and we can only find that out by going into the client companies and asking people what they thought of the programmes.

'If the programmes are being used for team briefings, we would ask the team leaders to monitor the effects, such as the levels of discussion which they stimulated. We are keen to show that we are willing to change to meet the needs of the audiences. If they are not going to be viewed in teams, we would send our business development people to visit places where the programmes are being shown and get them to ask questions.'

TWO-WAY COMMUNICATION

When talking about business television, we tend to think of it as a management tool for disseminating information, which means telling other people what to do in the nicest possible way.

The screen, however, should be useful for channelling information back up the other way, like a sophisticated suggestions box.

'Video can provide a "window on the world" for management,' says Stephen Batiste, chairman of a company called Commercial Presentations. 'We made a programme for senior management in British Rail, looking at how people at the bottom of the company viewed things. It was called *Where Are We Now?* and it had staff talking about their jobs. The managers were amazed by the perception and eloquence of their own people, people whom they would never have got round to talking to in the normal course of their work.'

A COMMUNICATIONS TOOL, NOT A BARRIER TO COMMUNICATION

The great danger with using television is that it will be seen as a negative force rather than a positive one. Employees who are shown a video of their chairman could react in two ways.

They could be interested to see and hear the message and feel pleased to have been given information that in other circumstances they might never have been privy to. Equally, however, they might see the video as confirmation that their chairman can't be bothered to come down and talk to them face to face.

Any senior managers who think that television will provide them with a way to escape the disciplines of 'management by walking around' are going to be creating exactly the wrong sort of atmosphere. Under those conditions television does become like something from an Orwellian nightmare, where leaders only appear on screens and the little people have no voices.

If it is to work, then it has to be part of a two-way communication. This can be achieved in many of the ways which we are talking about in this book, like talk-back facilities in live broadcasts, interactive videos, and training sessions where television is a tool for the presenter, not a replacement for him.

The mental leap which everyone has to make is away from the old-fashioned idea of television, fostered by the broadcasting organisations, that the people on the screen are the establishment talking to everyone else out there, telling them how things are going to be. Business television is merely one more channel of communication which can pass messages both up and down the line and back and forth between a company and the outside world.

WHAT IS TELEVISION GOOD AT?

Business television is not a panacea for all of a company's communications problems, and it certainly isn't a replacement for other forms of communication like print, slides or the spoken word.

It is just one more medium which will be suited to performing some of the communications functions.

'Television is good for giving audiences access to experiences they couldn't otherwise have,' says Michael Rodd. 'It can take people to locations and introduce them to subjects that they would not be able to reach in real life. Just as broadcast television can take viewers onto the pitch at a Cup final or the stage at Covent Garden, business television can take factory workers in Solihull into their company's head office in Neasden and vice versa. It can take people to the rain forests of South America, or down the salt mines of Eastern Europe.

'It widens people's perceptions and broadens their understanding of the world outside their own patch. It provides them with the ability to interpret things better.'

PRODUCT DEMONSTRATION, EXPERIENCE SHARING

'It can act as a product demonstrator,' Rodd continues. 'It could take viewers with the test driver of a new vehicle, for example, bringing them as close as is possible to experiencing what it is like actually to be there. It shares experiences with people in a way that it is not possible to achieve on paper. If several people in an organisation have been through some change process, for instance, they can explain to others about to undergo the same process what it was like. It might not be possible to get those people together to talk face to face in the same way. Just like broadcast television, it can be intriguing, exciting and stimulating.'

Television also offers the opportunity to reach large audiences very quickly with the same messages. However, a note of caution is sounded by Michael Rodd.

'In my opinion it is not good at putting across *detailed* information,' he says, 'like complex lists of figures. That sort of material is better on paper, where people can read it at their leisure, without being pushed on to the next image or idea before they are ready.'

Television can be useful for giving an idea of the climate and atmosphere of a company, a product or process. It will provide general impressions which can be backed up with specific information in other ways. It is an attention grabber and an illustrator.

Business television, in other words, has many of the attributes of broadcast television, and if used with care can add a few more.

Chapter 5

Private Corporate Networks

The biggest potential development in business television comes from the introduction of satellite transmission. It suddenly changes the way that company television programmes can be distributed, and it provides the possibility of two-way communication. It is all very different from the video in the plastic box sent through the postal service.

Those who work in this sector say that it is true 'business television' while the rest should go under other names, like 'corporate videos'. The terminology is confusing, but in the long run it is all an extension of the same idea.

The private networks revolve around the principle of broadcasting a single message or programme from a central point to any number of other sites. Telephone links provide the opportunity for viewers to respond and talk to the broadcasters live, although the broadcaster will not be able to see the people who are speaking to them. (As soon as you have both ends able to see one another you have moved into the area of video conferencing, which we will cover later.) It is the same principle as a 'phone in' on broadcast television, with the presenters talking down telephones to invisible but audible callers.

This is different from video conferencing, (see Chapter 8) which requires each site to be linked visually as well as by a direct sound line, so that the people at each end can see the people they are talking to as well as hear them. There is some overlap of terminology, however, when people set up ad hoc networks at conferences.

THE AMERICAN EXPERIENCE

In America, by 1990, there were said to be around 100 established private satellite networks (some estimates put the figure as high as 360), servicing 28,000 sites with 20,000 hours of programming for companies like General Motors, Ford, Wang and Texas Instruments. The average corporate network has 300 sites and broadcasts for 1 hr 20 mins each day.

Xerox in the USA educates over 10,000 employees via a dedicated network, while Domino Pizzas, which operates through a nationwide franchise system, uses a mobile studio with satellite facilities to make transmissions throughout the country. IBM has established three separate networks for corporate communications, training and customer liaison respectively.

Merrill Lynch produces three half-hour magazine programmes a week for transmission to their 600-odd sales offices. In 1984 they established a pilot network to fifteen of their offices. They believed that television programmes were more concise than written material and that video tape was too slow. In the first twelve months of the pilot study, the offices which had the network installed showed sales 40 per cent ahead of those without it. In 1988 they invited clients to view a programme broadcast across the US, and the following week their sales were up 20 per cent on the year's weekly average and 34 per cent on the previous week.

Federal Express transmit a five-minute summary of the day's activities live at the beginning of each shift change. It includes information on the current status of the company, such as how many parcels moved yesterday, routing problems due to weather conditions and information on new corporate clients won.

Computerland broadcast product reviews from their major suppliers, as well as service authorisation training, franchise communications and customer seminars through their network. Chrysler uses its network for parts and service training, marketing education, manager up-dates and dealer education.

Private networks have therefore been a reality in America for some years and some American companies such as Wang have even been beaming their corporate programmes across the Atlantic via Bright Star. Initially the high cost of bringing the programmes down to earth from the satellites, due to the charges levied by the various European telecommunications authorities, made this an impractical thing to do, but powerful developments in technology have meant that the programmes can be beamed direct to small satellites at company premises.

Close family relations of the corporate networks are the ad hoc networks created specifically for one-off events. These are discussed in Chapter 8, along with industry networks.

In the USA the latter are already up and running and are targeted at specific industry sectors such as automotive, medical, legal, financial and general management. The idea works along the lines of a trade magazine or newsletter, and provided the broadcasters can convince their potential audiences that the information which will be appearing will be vital to them professionally, they can get people watching for fear of missing something which the competition is tuned in to. There have been some attempts to set up industry networks in Europe, but so far most have stumbled.

It is in the corporate network field that the big potential growth exists, because it is here that the broadcasters, i.e. the companies, have a large degree of control over their audiences, i.e. the dealers or employees.

THE GLOBAL PLAYERS COME TO EUROPE

It is mainly the global corporate names, who are active in the USA, who have been looking at ways to do the same in Europe.

Progress in Europe during the Eighties, however, was slower than in the USA, largely due to legislative restraints. Unlike the USA, where the so-called 'Open Skies' policy means that private users are free to access video material direct from domestic satellites, in the majority of EEC states up- and down-linking has until recently been the exclusive province of individual telecommunications organisations. This makes satellite transmissions complicated and expensive to arrange.

Just as domestic air fares within Europe are sometimes higher than international fares taking you halfway across the world, satelliting around the relatively short distances of Europe was far more expensive than criss-crossing the whole of America with programmes. Liberalisation has been coming to Europe, however, along with a number of entrepreneurial companies anxious to help their clients overcome the various hurdles put in their way by bureaucracy. There are a lot of companies eager to start using the technology if they can just be shown a way of making it cost-effective.

THE ARRIVAL OF SATELLITE IN EUROPE

Things change fast in the satellite television world, and the merger between Sky and BSB in 1990 has meant that BSB Datavision's future, at the time of writing, is unknown. The services which it provided, however, will act as a prototype for the way things will go in the future, whatever the programming and broadcasting companies are called, and whoever they are owned by.

When BSB launched their five-channel network of subscription television, they invested many millions in expensive technology, including the launching of satellites and the establishing of ground stations. Their programmes were largely broadcast from lunch time to the early hours of the morning, leaving a major part of the day idle and unused. When the IBA gave out the franchises for satellite television, it stated that the companies must also be able to generate revenue from non-broadcast activities.

The specialism which BSB had to sell to the market was the ability to reach closed user groups with television programmes. It seemed a natural for business television. So BSB Datavision was born to market the broadcasting facilities of BSB to the business community for conferencing or instant and direct communication with employees, the press, or selected groups of customers or shareholders across Europe.

Security was achieved naturally from the fact that the BSB movie subscription company already required an encrypting (scrambling) system to ensure that no one who hadn't paid their subscription was able to receive the product; Datavision was thus able to offer the same service to business users. It is important, after all, that if one high street bank is broadcasting a series of motivational programmes to its people, the rival banks can't pick up the programmes simply by placing an aerial on their roofs.

Price Waterhouse Launch the Service

The BSB Datavision service was launched on 21 March 1990, with an inaugural programme from accountancy firm Price Waterhouse, analysing the Budget for its customers. The programme was made overnight at Price Waterhouse's head office in London, in conjunction with the *Financial Times*. It was half an hour long and was broadcast at 8.30 the following morning, at which time clients were invited to all the seven regional offices of the group in order to watch.

'It was a low-budget operation,' explains Neil Coles, commercial director of BSB Datavision, 'introduced and closed by Peter Hobday, with four or five experts talking against a simple background. It was a very effective demonstration of how the system could work.'

The Natwest Joins In

The National Westminster Bank was also quick to look at the possibilities of the system. The bank was already a major user of corporate television, making twenty or more videos a year and distributing them to their 3,000 branches for training and informational use. What they needed to find was a more cost-effective method of distribution which could cut out the messy and expensive business of copying the tapes.

Their plan is to send the programme down to the branches, where recorders will have been automatically set to record. There is a gadget which BSB Datavision developed to help people avoid the pitfalls of forgetting to turn the recorders on, or incorrectly programming them so that they end up recording something totally different.

Comet Links Up

Comet, the electrical retailing group, had different reasons for approaching BSB Datavision. They wanted to introduce a new initiative, broadcasting half-hour programmes to all their 250 sites on Saturday mornings.

They wanted to be able to invite their staff to come in half an hour earlier in order to see the programmes, which were designed to demonstrate that the company was at the leading edge of technology, while at the same time getting the managing director in front of the whole staff on a regular basis with messages about products and the company.

Dramatic Sales Increases at Radio Rentals

Radio Rentals went one stage further with BSB. During a national sales week they broadcast a daily programme live to all their 650 shops. It was broadcast from the studios of Wyvern Television in Swindon, one of the major corporate television production companies, for six mornings in a row.

The show was designed to work like a television telethon, and was renamed a 'Rentathon'. It was fronted by one of Radio Rentals' own regional sales managers and featured managers from the company talking about products and setting specific targets on each day. There were prizes to be won, names appeared on the screen and some of the winners were invited into the studio on the following days.

The whole show was designed along traditional motivational lines, but with the added dimension of being broadcast through the screens, making it unnecessary to take people away from their places of work.

'Radio Rentals believed they would be able to raise their revenue for the week by 50 per cent as a result of the programme,' says Coles. 'They actually managed to raise it to 300 per cent of a normal week's revenue. Which means that they effectively added two weeks' revenue to the bottom line for the year.'

The Biggest UK Network for BMW

BMW are the first company in Europe to set up a full-scale private network. They are now broadcasting regularly to their dealerships using the communications consultancy Visage to make the programmes.

Each of the 160 dealerships has a dish on the roof and programmes are beamed down containing technical information for the service staff, sales and marketing information on what the competition are up to and anything else which will be of interest to the staff.

Each programme is broadcast from one dealer showroom to the others, to give everyone a chance to appear and to see what others in the family are up to. When there is an event like the Motor Show, the programme comes live from the BMW stand. Each show is a mixture of live programming and filmed inserts, just like broadcast news programmes. Although the programme has a television professional acting as a link, the people who appear are from the company.

By making the audience part of the broadcasting team and using familiar faces, you avoid the danger of the dealerships feeling isolated and 'talked at' rather than 'communicated with'.

'Although BMW is the first company actually to take the plunge over here,' says Bob Clarke, chairman of Visage, 'a number of others are showing an interest. Their first reaction when we approach them tends to be, "Business television? Never heard of it. Sounds expensive". When we tell them exactly what is involved they tend to change their tune. "Incredible" is a common reaction, "will it work for us?".'

According to *Car* magazine there is a document within BMW called 'The Vision', part of which says that the company and its dealers will function better if individual initiative, instead of corporate edict, is allowed to thrive. To do that requires changes in the way the company and dealerships are run. Improving communications is a cornerstone of 'The Vision'. A private network allows everyone in the dealership to see the same information, presented in the same way. No longer will staff be briefed by various levels of management who, presumably, each add intonations, inflection and interpretations to the messages

'The sort of companies which are installing private networks tend to be pretty sharp communicators anyway,' says Visage's Bob Clarke. 'They are already good at prioritising information. All we do is make it happen for them. As the BMW network establishes itself, it will become more proactive.

'There will be rolling programmes of messages and sequences of events, things that will draw the viewer back for more, just like in broadcast television. With a continuous channel of communication like this, you can keep talking and developing stories, and you can do things like breaking bad news gently. To make these sorts of programmes, a production house needs to be able to understand the corporate mentality, and also to be experienced at making long-running, low-cost programming.

'Research in America shows that people in companies which have private networks are viewing an hour or more a day of programmes. DHL, the giant international courier company, is so convinced of the value of it that they have their own satellite which they lease out to other people.

'There is a belief that you need a certain number of outlets to make a private network worthwhile, but that is not true. What matters is the relative importance of the audience and the urgency of the information.

'If you are a merchant bank and you have ten major clients, who are the ten most powerful fund managers in the City, it would be worth having a network with ten outlets which reached these vital people, if that meant you could communicate with them without dragging them out to a hotel somewhere to watch a slideshow.'

THE BENEFITS

'The benefits of private networks are that they are very accessible, very immediate, very consistent and very secure,' Clarke explains. 'You can reach people and motivate them in a very immediate way. Next to face-to-face meetings, it is the best way to communicate. There are also cost benefits when you start to work out the number of different things which you can do on them.

'When the Chancellor announced Britain's entry to the ERM and the consequent drop in interest rates, we made a programme for BMW to send out the next day, in which we talked to the financial people and senior managers to find out what this all meant to the people doing the selling. We were able to talk to the top people in the company and ask them the questions which the audience wanted to ask. There was no other way we could have got that information out to so many people so fast.

'It comes down to the difference between hearing and seeing a director explaining something, and reading a memo on the noticeboard with the director's signature at the bottom. One is personal and one isn't, one allows the viewer to form judgements of the person, the other doesn't.

'People don't have to be limited to watching the programme at the time it is broadcast. Some programmes can be repeated, some can be recorded and shown later, and some can be aimed at specific sectors of the company. One programme might be aimed at the technicians, another at the sales staff.

'It is just the same as family viewing in the home; there are some programmes for the children and some for the parents.'

BEAMING-IN PROGRAMMES FROM TEXAS

Texas Instruments, the high technology giant based in Dallas, has been a heavy user of its private network in the States for several years, and in 1990 it started broadcasting to all its manufacturing sites in Europe. The network was set up for them by SMI, a communications company owned by British Aerospace.

The programming comes in from Dallas, but it is all interactive, so that viewers at any of the sites can talk back to the broadcasters, making the material both personal and relevant to their situations. The programming is covering everything from technology interchange material to sales information. In America the company broadcasts several hours a day, and not all the programmes will be suitable for European transmission initially.

'There are a number of American companies following these sorts of initiatives in Europe,' says Peter Thompson, chairman of SMI. 'Many of them are in the high technology field, others are in the automotive industry, financial services and the medical world. The banks and insurance companies are all doing large-scale feasibility studies prior to investing.'

Not so long ago the idea of a private, corporate television station would have sounded like something from a science fiction novel, perhaps a bit threatening, like the onset of Orwell's dreaded *1984*.

Now, however, they are a reality, and before long they will be commonplace in any company of any size. What makes it less threatening to those who have used it is the fact that it can be a two-way experience. Just as the fears that computers would depersonalise us have been proved to be unfounded, and in fact skilful database manipulation has increased the personalisation of many services, well-made business television will bring people closer together rather than coming between them.

It will improve communications between people, cutting out middlemen and the more impersonal and old-fashioned means of communication.

STAFF TRAINING

'At the moment,' says Graham Higginson of Blackrod, 'many multiple retailers close their premises for half an hour a week in order to undergo staff training. This is often being carried out by a junior member of management who may, or may not, be a good trainer. There is little chance that the messages being put across in all these stores are consistent with the company's original philosophy.

'If, on the other hand, they were all sitting down to a training programme being broadcast from the company's head office, asking questions down the telephone lines and being given direct answers, you have a training programme of an altogether different calibre.

'Imagine also the motivational power of a company's chief executive addressing the staff live when something good happens, or the impact control if something untoward occurs in the company.

'At the time of the 1987 stock crash, Merrill Lynch successfully used their network to sustain confidence in their salesforce and guide them through how to deal with their panicking clients. They claimed that that one programme made their entire investment in business television worthwhile.'

LINKING WIDESPREAD ORGANISATIONS

A private network would work for any corporation with employees scattered in a number of different locations. It might be in the retailing industry, with stores dotted around the country, or around the globe, or it might be a major manufacturing company with widespread offices and factories.

Imagine, for instance, a chain of fashion retailers with a good database marketing system set up to tell them which of their lines are selling best at any given moment. With the help of a television network, they could tell every store on a Friday how to dress their windows to make the maximum impact on the marketplace over the weekend. They could actually show them pictures of how the windows should look, and answer any questions they might have which are specific to any one particular store or geographical area.

They could, in other words, react immediately to the demands and whims of the marketplace, maximising the performance of all their stores simultaneously.

Perhaps in the not too distant future every large company will have a permanent broadcasting facility, on the same basis as the Radio Rentals experiment but run every day and all day. In every communal area around the companies there will be large, perhaps even high definition, television sets. They will stand in canteens and recreation areas, next to coffee machines and in reception areas.

Each day the companies will provide a mixture of information for different groups of employees and opportunities for two-way communication. They might want to explain a new management decision or bad results to them. They might want to announce the opening of a new store or factory, or explain about a bad story which has appeared about them in that day's papers. They might want to show managers how products are made in another part of the world.

Films about every part of a company's operation could be shown to people in the company who would otherwise have no way of seeing the broader perspective, or of knowing how their job slots into the overall picture.

On top of that the company might want to distribute information concerning product or price changes, or personnel information about pay or pensions, or healthcare announcements. They could do profiles of employees about to retire, or play staff videos of weddings or parties.

The unions, or the company sports clubs might want to make announcements and to explain aspects of their work. There could be daily phone-ins for people to question members of management or the unions about anything which is worrying them, and anyone with a genuine problem will have open lines of communication to the people who could solve that problem.

Barriers, misunderstandings and bureaucracy begin to break down as soon as people are talking directly to one another; even if they have to do it via a satellite.

Then there could be training programmes which the personnel departments want to put over to all or some of the employees at certain times of the day, or case studies of the companies' biggest clients, which the public relations people think would boost morale and improve customer care understanding.

A corporate television network can serve a corporate community in just the same way as broadcast television serves a local or national community, with a mixture of information and entertainment, news and education.

ALLOWING DATA TRANSMISSION

One of the reasons why America is so far ahead of Europe in private networking is that they are allowed to use satellites for data transmission. This freedom is only just coming to Europe, but as it comes it makes the economics of private networks completely different.

It makes the idea far more acceptable to large organisations who can easily understand the benefits of being able to transmit data around the globe, and can see how the same networks could also be used to send television programmes very cost-effectively.

HOW DO YOU START?

The first move is to get a company in to do a feasibility study, which means looking at your communications needs and the methods which you are currently employing.

It may be that television would be completely unsuitable for your company. It could also be that it would offer you cost savings by rationalising other media. If you are producing a number of glossy printed publications, for instance, and organising expensive roadshows and conferences to put across a message which might as easily be conveyed through the screen, a television network could be a highly cost-effective alternative.

You can then research the market further by making a pilot programme and trying it out on the potential audience. You might start out using professional broadcasters to front the pilot, but if it became a regular programme they could be replaced by people from within the company, who could be trained in the same way as someone in the public relations department can be trained to edit the company magazine to professional standards. The programmes could be made either in-house or by outside production companies.

If the pilot proves to be a success, the company can then start to make programmes regularly, which may mean every month, week or even day. To save costs, these programmes can be sent down from the satellites at off-peak hours (just like off-peak electricity or telephone calls), and video recording machines can be set up at each location to record them so that they can be replayed regularly, (or even constantly), during the following day.

THE POTENTIAL

The long-term potential for these sorts of networks is enormous. Imagine the problems of a motor manufacturer wanting to get across a message to mechanics about a new spare part.

To hold training sessions, or to send demonstrators round to every garage using their products would take months or even years, and the chances of persuading mechanics to sit down long enough to read a leaflet are very slim. But a television programme, flickering away beside their coffee machines, designed to catch and hold their attention, is going to get the message across quickly and effectively.

Once the system is installed it can then be used for other applications. Good customers of this garage chain, for instance, could be invited in to see advance programmes about new models not yet released to the media or general public. At best this could result in actual orders, and at least it would help to engender good public relations.

Allowing the audiences to talk back to the broadcasters over the telephone lines is vital if a company genuinely wants to communicate with its audiences, since one-way communication is always going to be inferior to two-way.

If, for instance, a company's chief executive is announcing the annual results, or explaining why the company is laying people off or closing a factory, it might be impossible for someone that senior to visit each site and explain in person what is going on. Television will provide the next best option, while giving people at each location a chance to answer back and ask questions.

NEVER REPLACE FACE-TO-FACE

The idea of a massed workforce talking to their leaders on a screen does bring us back to the imaginings of George Orwell and *1984*.

Of course it is impersonal, and is not as good as meeting the people face-to face in order to discuss the issues. But in a large company, what chance do most of the workforce ever have of meeting and questioning senior management in any other way? In many companies the people on the factory floor won't even know who sits on the board of directors. At least by seeing them regularly on television they can form some sort of opinion about them, and if they are communicating well they will be able to engender feelings of loyalty and motivation in the hearts and minds of their audiences.

Television should never be used as a way of distancing people who are currently working together closely. It can, however, be used to bring together, if only electronically, people who would normally never have a chance to meet and talk.

HOW MUCH DO PRIVATE NETWORKS COST?

The costs fall into three main areas.

- First there is the capital cost of the reception equipment, which could be anything from a few hundred to a few thousand pounds per site, and might involve some annual maintenance costs on top.
- Then there are the transmission costs which are charged strictly on the amount of time you use.
- Finally there are the programming costs which will depend on the production values of the programmes, but which will be considerably lower than for the production of comparable videos because of the savings made through planned scheduling as opposed to ad hoc purchases.

Promoters of corporate networks claim that on a 'cost per hour basis' a broadcast via business television works out at 30 per cent as much as a conference bringing together 1,200 people. In terms of 'cost per person, per time', however, they claim business television costs 4 per cent of the price of the same sample conference.

Although the installation of a private network is a major financial commitment for an organisation, the costs of producing regular programmes over the air is not as high as might be supposed.

Neil Coles of BSB Datavision says, 'When you make a one-off video it is bound to be expensive, because most of the costs are at the production end. You are using everything once and then throwing it away. If you are doing broadcasts every week or month, however, you can use the same studio set-ups and begin to bring prices down.

'Companies who are currently distributing their programmes on videos, as the NatWest were, will find that using satellite provides considerable cost savings, particularly if they are currently using high standard tapes and making that many copies.'

BSB Datavision claim that a company communicating with 200 sites or branches should find that the costs of satellite would about break even with the cost of producing cassettes. However, that is before they have taken into account all the new things which they will be able to do once the network is set up, which were impossible before.

'The real value of services like ours,' says Coles, 'will come to be appreciated when companies grow used to regular televisual communication. When employees of a company have grown used to seeing senior managers talking live, and have become accustomed to talking to them on air over the phone lines, they are never going to want to go back to the bad old days when senior managers were names at the bottom of memos and pictures in company magazines.

'Once management have grown used to being able to broadcast training programmes whenever they feel the need, and to inform the whole company simultaneously about changes in policy or new product launches, they will not be able to give such luxuries up.'

Globalisation was a buzz word which gained a lot of popularity in the Eighties. In the Nineties the moves towards a unified Europe are again focusing people's minds on the idea of one-world, and the Global Village which Alvin Toffler talked about in his book *Future Shock*.

The satellite and telecommunication technologies largely do away with borders, apart from the need to deal with different post offices and telecommunications authorities within Europe, which is still difficult at the time of writing, but not impossible.

CROSSING LANGUAGE AND CULTURAL BARRIERS

The question is, do your messages easily cross language and cultural barriers? In some cases English is the business language of the companies which are involved in widespread television networking, which means that they probably 'get away' with talking to all their international people in one language. But 'getting away with it' is not the most effective way to communicate.

It is possible to translate videos in the same way as you can translate printed literature, provided you have not filmed a presenter on camera who cannot then be dubbed without looking ridiculous. Simply recording a voice-over in a different language is not too expensive, but if a programme involves a great many graphics and bar charts, and they all need to be translated into Japanese, the production costs will begin to soar.

It is quite possible to employ simultaneous translation techniques when broadcasting live to multiple points, just as you would in a conference hall if the audience was made up of a mix of nationalities.

When computer company Digital sent out a broadcast from England to thirty-four different offices and meeting places around Europe, they had translators in five different languages working on the project.

Maxwell Michie of Shell says, 'Whenever we are making programmes we have to bear in mind the fact that they could be watched by a man in Brazil, a girl in Singapore and a class of students in Nigeria.

'There are certain things which do not translate well internationally. Humour is one example. We also try to avoid drama because if the film is going to be shown in a number of different countries you need to have a mixture of races represented so that no one feels alienated. The result can appear very artificial.'

Language isn't the only communication barrier. An international company may be based in America, but it doesn't mean that the bulk of the workforces in Britain or Australia share all the cultural values of their parent company. Motivational techniques, for instance, which might work well in the States, could appear laughable to an English audience.

As long as international broadcasting is used as a two-way communication this is not a problem, but if it is used simply as an informational tool, the receiving countries may grow to resent the imperialist feel of being talked to from another country.

Chapter 6

The Corporate Video

This is where most companies start with television. They decide that the time has come to 'have a corporate video', just as they 'have a corporate brochure'. In fact they probably expect initially to be able to make a 'film of the book', and simply transfer the contents of the brochure into moving images.

Although it is the way most people start, it is, by and large, a mistake. Television is not primarily a verbal medium; it is a visual one, albeit with words to back up the pictures.

If a company makes a video for a special reason – recruitment for the university milk-round, a programme to show to the investment community prior to a flotation or when funds need to be raised, a motivational programme for a company conference, or an introduction to a company for showing on an exhibition stand – the programme might also prove to have a more general appeal as a corporate video.

It is important, however, that the specific need is recognised first, otherwise the programme will lack direction and will end up doing no particular job.

WHAT'S IT FOR?

There could of course be a number of good reasons for making a straight corporate video. It might, for instance, be needed purely as a means of putting across information about a company in the same way as a brochure, but to audiences who might be persuaded to watch a screen more readily than to read a brochure.

It can, in those cases, work in tandem with a brochure, but it must be used for its intrinsic visual advantages if opportunities are not be to be missed.

There are many ways in which the visual abilities of television can put across messages which cannot be conveyed by written or spoken words alone. We only have to think of the ability of the pop videos to sell records to see how powerful a moving image on the screen can be when it hits the right response buttons in the viewer, catching their attention, making clear messages that might have been difficult to hear on the soundtrack alone, adding new imagery which audiences might not have thought of for themselves, and providing an entertainment and stimulation of its own.

Employee Relations

One of the most effective uses of corporate video has been in the area of employee relations. Many companies have been producing regular employee information programmes for many years (World Wide Pictures has been making a quarterly programme called *Pipeline* for BP since the late Seventies), and one-off programmes have often been used successfully to handle a particular problem of human resources or industrial relations.

Everyone is aware of the game Chinese Whispers, in which one person whispers a message to another, who whispers what they think they have heard to another, and so on down the line until the final message received bears no relation to the one which was originated. In the industrial relations arena, with so many emotional issues and vested interests at play, the rumours game can lead to catastrophic results.

'When the Mersey Docks and Harbour Company wanted to reduce their workforce from 13,000 to 1,500,' says Ray Townsend of World Wide, 'they wanted to ensure that the same message got through to everyone, and that there was no room for adjustment along the way.

'They also offered the unions the right to talk to the camera, and did not censor what they said. What they were after was a consistency of message. They wanted to make sure that the union leaders were saying the same things to their members as the company were saying to their own managers. A lot of industrial relations problems arise out of simple ignorance, managements simply do not communicate in a direct way with their workforces. Even with a militant workforce it is possible to achieve a high level of understanding if you talk to them directly and honestly.

'Television is also very effective at revealing hypocrisy and cant, particularly if you print up the words being spoken and distribute them verbatim to the target audience. Anyone who is working from obvious self-interest, and talking illogically or in clichés, will soon be exposed, whether they are from management or the unions.'

Conveying the Feel of a Company

Television pictures can do just the same for a company, conveying a glamour and excitement which might not be obvious to anyone looking at the company from the outside, or even to those working within it.

Most companies can provide a mixture of satisfactions to the people they employ and the customers they serve. They can offer interest, companionship, security, team spirit, tension and excitement, pride in the product or service, and many other intangible things. A creatively conceived programme can pick out what these emotional responses should be in the audiences and then seek to satisfy them by showing the company from different perspectives.

If, for instance, it is to be a recruitment video, then it will need to convey the atmosphere and culture of a company in a way that the printed word cannot.

It is almost impossible to describe the attributes of a good employer in print without resorting to the banal, cliché language which is used by every company, regardless of whether the words are true or not. With television, however, you have an opportunity to illustrate what it feels like to work for the company.

Whether it is by being a fly on the wall watching employees at work and play, or whether it is through vox pop interviews with members of the staff, it is all going to be more believable and accessible than some carefully worded and edited piece of literature which has obviously emanated from the minds of senior management.

Giving the Whole Picture

Someone working as a parcel sorter in an airport warehouse might not be able to see anything very exciting or glamorous in what they do. A corporate video made by their company, however, could show the global picture, giving a sense of the urgency for a parcel to reach its destination on the other side of the world, perhaps in order to save a life or to make a crucial sale. It can show the immense feat of organisation and high technology needed to integrate motorbike messengers, lorries and vans with the aircraft which link up the various national arms of the operation.

Suddenly the parcel sorters are able to see themselves as part of a giant international network, part of something important and exciting. Short of taking them all round the world and introducing them personally to everyone in the chain, a television programme is the most effective way of conveying that excitement, and giving the audience a broad picture of where their jobs fit in.

Annual Report and Accounts

A video can also be a useful adjunct to the annual report and accounts, to be used at the annual general meeting in order to explain what is happening, although as we have said it should not be used to convey detailed financial information which is better suited to print.

What better way to counteract the impression given by a set of accounts which have failed to live up to the expectations of investors, than to show film of the company in action, installing new machinery, making products, meeting orders, looking after their people, selling and advertising their goods?

Marks and Spencer make a video each year to go with their end-of-year results. Initially it was designed for the financial community, to show the City a summary of the previous year's achievements. The same basic material, however, is then turned into a staff video, translated into all the necessary languages and sent to each store, giving everyone an idea of what is happening in other parts of the company, particularly other countries.

The same programme is also circulated to new suppliers, export customers, franchise operators, and anyone else who needs to have an overview of the company's activities at that moment.

M&S also have a six-minute corporate film which charts the company's rise from a penny bazaar to the international giant it is today.

Conveying a Spirit of Strength and Optimism

In a video, industry experts can be seen talking about how they view the long-term prospects of the company and its sector, conveying optimism and confidence in a way that would never be possible on paper.

In some cases companies have to rely heavily on the building of a mystique in order to differentiate themselves from their competitors. The airline industry is a particularly good example. All the big national carriers offer much the same services, so to stand out from the crowd they need to build up an image, and this can be done highly effectively through the use of film and television.

When GHA made a corporate video for Japan Airlines, for instance, they explored the Japanese character through British eyes, thereby crediting the airline with the same strengths as were perceived to belong to the Japanese people.

They showed the formality, ritual and hospitality of the tea ceremony, where calm and courtesy rule; they showed a modern Japanese car factory, with robots working cleanly, efficiently and reliably, and they drew parallels with the way the airline was run.

Singapore Airlines have been the most dramatically successful in this area, their image being built on the beauty of the stewardesses and the mystery of the Orient, coupled with the reputation which Singapore itself has for cleanliness, efficiency, modernity and luxurious hospitality.

When it comes to finding new ways to put across old messages, television offers a host of creative possibilities.

VIDEOS ARE NOT THE SAME AS COMMERCIALS

'The danger is that companies will try to make their corporate videos look like their advertisements,' says Roddy Gye of GHA. 'They try to write twelve-minute commercials. The problem there is that a commercial only runs for thirty seconds. Something longer is going to need several changes of pace if you are to hold the concentration of the audience. You need a mixture of ingredients like story-telling, enigma, surprise and humour.

'To get the right balance of these ingredients, you have to understand the needs of the audience, their aspirations and the circumstances in which they are viewing the programme.

'It is always better to go to the production company with the problem, not with a half-formed idea of the solution. Otherwise you will be choosing the production company whose suggestions conform most closely to your own ideas, whereas in fact you might need something completely different, which will only emerge when the problem is viewed thoroughly from an outside perspective.'

Before you make a corporate video, therefore, be sure you know exactly what you want to achieve with it in the short-term, and then take any long-term usage as a bonus.

INVOLVE THE EXPERTS EARLY

'A common mistake which clients make,' says Michael Rodd of Blackrod, 'is to come to us when they are much further down the line in conceptual terms than they should be. They often don't understand the level of service which they can expect from organisations like ours. In some cases they even believe they have to write their own scripts. Yet they would never expect to walk into an advertising agency with a finished script for a commercial, or into a design consultancy with proofs of the packaging they want to commission.

'Some production houses which are keen to make money at any cost will happily take the scripts they are given and work with them, when in fact they should be educating the clients on how to use the services of professionals.

'Too many production companies in the past have been too keen to build up their bank balances fast, and that has led to many unhappy experiences for clients.'

TOTAL QUALITY PROGRAMMES

World Wide Pictures has done a great deal of work with companies instigating Total Quality Programmes such as Leyland Daf and BP Chemicals, using television to spread and illustrate the messages to employees at all levels. Total Quality Programmes are an approach to improving the effectiveness and flexibility of businesses as a whole. They are a quality process which involves every person in every department of an organisation working together to eliminate errors and prevent waste. The benefits of making sure things are done right first time include greater efficiencies, lower costs, improved quality, better reputation and greater market share. To make it happen, however, you have to be able to communicate fast and effectively with everyone within an organisation, which provides an ideal opportunity for video.

'The good thing about quality programmes,' says Ray Townsend, 'is that the results are measureable, which means you can justify the programmes and their costs. As soon as you can prove to management that something has worked, it is not hard to convince them to repeat the exercise. For a company like ours to thrive we need to have a core of clients coming to us with repeat business.'

UNDERESTIMATING THE AUDIENCE

'The most common mistake which managements make in the UK is to underestimate their workforces,' says Townsend. 'They try to get away with saying things to camera which they would never dare to say to anyone face-to-face.'

If they were talking to someone direct they would have to address the issues which were of most interest to the listener, because that is what the questions would be about. Where the audience primarily wants to know whether their jobs are safe, there is no point making a video which talks about a thousand other things but doesn't answer that question. As long as it is the issue which is at the front of the audience's mind they will not be interested in listening to anything else, and they will be suspicious of anyone who is not addressing the key issue.

Management may think they have made a clever programme, that they have managed to evade contentious issues and smooth things over, but in fact they will have had no impact or effect on their target audience, apart from proving that they are out of touch with their own people and suggesting that they have something to hide.

'Never make videos for the sake of it,' advises Townsend. 'Only make them to deliver hard information.'

BEARING BAD NEWS

When a client wanted to achieve a two-year pay deal which was under the level of inflation, in an industry where neither practice was traditionally considered acceptable, World Wide created a programme which was the central part of a broad compaign.

The production team talked to as many people as possible, to find out exactly what the main concerns of the target audience would be, and what messages would make them sit up and listen. They then put together a package.

The company's management had a reputation for straight talking, but it was felt that an even more convincing case could be made if respected external commentators could be heard talking about the future prospects for the industry which the company was operating in. Although things were not bad at the time of making the programme, it was generally agreed that the industry was in for a downturn. Experts from the financial and academic worlds were therefore invited to put the case on camera as to why bad times were just around the corner, and how only the strongest would survive.

The management was then able to put forward their case for how they wanted to make the company strong and ensure its survival into the future. They explained how they needed to create funds which could be channelled into research and development, and did not try to pull the wool over anyone's eyes.

'They had truth on their side,' explains Townsend, 'and that shone through on the screen.'

VIDEO PUBLISHING

By video publishing, I mean companies that actually sell or hire out video programmes as products.

It might be a video on how to service a car or how to wind surf, it could be the management training films starring John Cleese, or a magazine programme about developments in the world of high technology. The common link is that the video project itself is devised to make money. If as a by-product it also promotes the name of the makers and enables them to sell other products or services, that is a bonus.

If, therefore, you are working for a management consultancy, you might decide that you can package one of your services and encapsulate it on video, so that a client could listen to you talking, or see you in action on a case study, and be able to understand how they could apply the same techniques to their company.

In the past you might have decided to write a book on the subject, in which case you would have been paid money by a publisher (if you were successful in placing it) and earned royalties on the number of books sold. On top of the money, however, you would have gained the prestige of being published and would have promoted the other services and products of your company by association.

The same applies to videos, except that there is not the same centuries-old structure of publishers, booksellers and direct marketing companies waiting for you to feed your ideas into their machine.

If you want to sell this sort of idea through the medium of television you are going to be largely on your own, unless you can fit in with one of the companies offering generic training programmes. Such companies are the closest thing to book publishers in the television world. There is no reason, however, if you have a strong database of names who might be interested in hearing what you have to say, why you shouldn't be able to make a programme which they will want to buy. Management consultancy is a field which is already well exploited in this way, through the traditional firms which market videos from the likes of Tom Peters and Michael Porter.

The sporting sector is also developing its market with everything from how to wind surf to improving your golf swing. There is no reason why companies that have been thinking about sponsoring broadcast programmes (which we discussed in the previous chapter) should not be putting their money into products which they can sell direct to the public.

A petfood company could make and sell programmes on pet care, a building society or bank could make programmes on how to plan your finances, a nappy manufacturer could produce a baby-care programme . . . There is no end to the possibilities.

Chapter 7

Television at Conferences

When Powergen was being privatised, the senior management wanted to hold a conference which would involve all 15,000 of their staff, so that they could explain what it meant to them. They wanted to allay the fears and insecurities of people who had worked all their lives in the safe environs of a nationalised industry, and the senior executives wanted to be able to answer their questions directly.

Naturally all fifteen nation-wide power stations could not be closed for a day in order to invite everyone to an event, and it would have been logistically impossible to move so many people to one venue anyway. The answer to the problem lay in taking the conference to the people. It lay in television.

The event was masterminded by Spectrum Communications, one of London's leading conference organising companies, and it was decided to stage a presentation at the National Exhibition Centre in Birmingham for 2,000 people from Powergen's head office. The senior executives would appear live on stage in front of this audience. A television link-up with the power stations was then organised and events were arranged at each location – ranging from something in the staff canteen to a marquee on the lawn, depending on the numbers of people who had to be accommodated.

At a pre-arranged time the presenters at each of the venues linked up with Birmingham over the television sets, and everyone was able to watch the presentation live. At the end of the 'show', they were then linked by British Telecom audio lines so that people at each site could ask questions of the executives.

The whole show had to be co-ordinated rather like a Eurovision song contest, moving from station to station for the different questions. One person at each location had been nominated as spokesperson, and the executives had been notified of the questions in advance so that they could prepare useful answers.

'We used Creative Technology,' explained Martin Pentony-Woolwich, a director at Spectrum, 'who organised everything with the satellite company, which was British Aerospace, and with British Telecom.

'The best use of satellite video is when a limited number of experts need to talk to a large number of others with interactivity. Some companies just use the television cameras as a sort of hype, but that is worthless unless the people at the other end of the transmission participate. By using television technology wisely you can save the valuable time of senior executives and achieve feats of communication which would simply not be possible any other way.'

The show was broadcast live, but it was recorded and edited and video tapes were sent out so that anyone who had missed it would get a second chance to hear the key messages.

'Whenever we use cameras at an event, we record it,' says Pentony-Woolwich, 'because there is usually some way in which a video can be used afterwards, and it helps to amortise the costs of the day.'

Spectrum did the same for IBM when they wanted to hold a two-centre product launch, in London and Manchester, to save bringing everyone to London. They hired the Royalty Theatre and a conference venue in Manchester. The show was staged in London, with an identical staging in Manchester but with the main event being televised from the London theatre.

An extravagant product launch by ICL was recorded on three cameras, edited to get the best shots, and the video was then distributed worldwide to be used as a presentation aid by their sales force.

'Conferences are using video more and more,' says Pentony-Woolwich, 'because of the influence which television has in all our lives. Because Britain has the best broadcast television in the world – or so we are told – the audiences expect high standards. If something is tacky they will recognise it as such immediately. Not only are they used to high standards of actual programme making, they also expect high standards of graphics, and all the other things which they are used to seeing on the screens in their living rooms.

'There will, however, be a place for audio-visual slides for a long time. You can do much more complex, intense and dramatic things with slides, and attain a quality of picture which is impossible on television, at least until the technology of high definition has come a lot further. They are also cheaper to produce, and simpler to use, not needing armies of technicians, which makes them more useful for small meetings. What you don't get with slides is the fluidity of movement, and anyone who puts slides onto video for a show is getting the worst of both worlds.'

CONFERENCES – A MAJOR MARKET FOR TELEVISION

Planning and designing conferences is a big business, particularly at the glamorous 'product launch' end of the market, and video is just one of the potential visual aids which can be used.

The main opposition is the traditional slide show, which provides the slickest and most professional pictures obtainable, and also makes possible a number of multiple images which would be impossible to achieve with video alone. Without high definition television, it is not possible to achieve such high quality pictures on a big screen with video. Television does, however, have other advantages, apart from the ability to have moving images.

If you are staging a major show at one of the giant conference venues, and you want to be able to show the material to a wider audience as part of an on-going marketing strategy, you need a medium which is easy to distribute.

It would be very hard to scale down a multi-slide show at the NEC and show it to someone on a desk-top monitor in Basingstoke. A video, however, can become a televisual record of the event, and can be edited to contain any sales message which is appropriate for other situations.

BEING IN TWO PLACES AT ONCE – OR THREE, OR FOUR . . .

The chairman of a large company was due to make a presentation to the media at a press conference in London's Savoy Hotel. A few days before, when it was far too late to cancel the event, he realised that he was not going to be able to get back from a business trip to Germany. One way out of the problem could have been to have someone else deliver his speech, but that could have seemed to the media that they were not considered important enough to merit the top man's time, and they might also not be interested in quoting a spokesperson when they had been under the impression that they were going to be able to question the top man.

The alternative, and the answer which this particular company chose, was to organise a two-way television link.

The audience were able to see the chairman and question him. He was able to answer their questions direct and explain why he couldn't be there in person. He showed that he considered them important enough to go to the trouble of arranging a satellite link, and gave an added buzz of excitement to the proceedings.

TELEVISION AT EXHIBITIONS

It is very tempting to use television on an exhibition stand, and very easy to waste the opportunities which it offers.

When planning a stand, or when facing the prospect of being stuck on one for a week, the idea of having some screens flickering away with the corporate message and information about the products is very tempting. It should add some life to the stand, brighten it up and give you something to look at during the quiet moments.

But it should also act as an attraction, luring people up out of the aisles, and should then do some of the initial presenting work, leaving the staff free to concentrate on anyone who is genuinely interested and passes the first phase of information.

In reality it doesn't always work that way. Once an exhibition hall has woken up and the people begin to flock in, there is usually a lot of ambient noise and many distractions. It is not an atmosphere in which passive viewing is going to be easy.

Secondly, people generally have a lot to do during their time in an exhibition hall and so they do not have time to waste. Most video programmes are linear in design, they have a beginning, a middle and an end. The chances of someone arriving at the beginning and staying to the end are very slight. It is also unlikely that anyone arriving in the middle is going to stand around waiting for the programme to begin again.

Television, however, does have some useful features at exhibitions if it is designed right. It can be an excellent image maker. The car manufacturers are particularly adept at this, using the screens almost as moving wallpaper, to put across images of beautiful machinery and macho driving.

If the film-making is glamorous and exciting it will bring people up to watch, for the sheer pleasure of it, and it will put across the right messages about the product without the watcher having to follow a linear story.

Another way of using the screen effectively is to ensure that the video consists of short, punchy sales messages, almost like advertisements in a commercial break on broadcast television.

If each message is delivered with enough impact, it will draw the audience up from the aisles and then it is up to each subsequent message to hold them there. Interactive screens may also be useful if a great deal of technical information has to be put across, allowing the visitors to the stand to set their own agendas and follow their own interests.

TALKING DIRECT TO THE EXPERTS

One of the frustrations for people visiting exhibition stands can be the lack of anyone with in-depth knowledge of the product or service on display. Few people enjoy waiting around on stands on the off chance of a customer coming along, and so the job nearly always falls to the sales team, who then have to refer back to their head office if they want to have any technical questions answered.

In an ideal world there would always be someone on every exhibition stand capable of answering any question that might be put by visitors, and of speaking authoritatively on behalf of the exhibiting company's senior executives and technicians. In reality this does not often happen. While the staff who are on the stand will be able to provide a first resort when answering questions, the use of video can provide another potentially powerful sales tool by bringing the big guns in.

From the point of view of the exhibiting companies, it would be completely impractical to have their best technicians or senior directors spending days on end attending exhibitions in foreign cities when they should be doing the jobs they are hired for.

What can be done, however, is to set up on the exhibition stand a small studio with a satellite link. Each day, at an appointed time, the video and telephone lines can be opened and the directors or technicians can be sitting in a studio at the company giving presentations and answering questions from the people who are visiting the stand. Another example of how television can be used to break down barriers between people, bringing them into direct contact more easily and more quickly.

Not only does this put the two sets of people at either end of the business relationship into direct communication with one another, cutting out the delays inevitably caused by corporate bureaucracy, it also provides an event for the sales people to build a promotion around at the point of sale. Anyone who comes onto the stand and shows an interest can be invited back to ask questions or listen to the technicians talking at the appointed time.

Without that communication tool the procedure of going back to the company after an exhibition and then sending out information, or re-contacting the people who showed an interest, can take weeks or months, and in some cases might not even happen at all, due to pressures of day-to-day firefighting.

TELEVISION AS A PERSONAL SPEAKER SUPPORT AID

The amalgamation of television with computing has meant that video graphics are now starting to replace traditional forms of speaker support aids like the overhead projector.

'We can provide speakers with whatever they want in the way of pre-production work,' says Graham Andrews of Creative Technology, 'whether it is bar codes or transparencies, a normal television set or a six-foot screen, and we can produce graphics which are the same standard as the sort of things that would appear on broadcast programmes like *Newsnight* and *News at Ten.*

'We can provide back-up staff who will take care of everything so that all the speaker has to do is speak, knowing that the right pictures are coming up on the screen at the right times. It cuts out the amateurish fumbling about for foils, or trying to find the right slide at the right time, and gives the speaker more opportunities for ad libbing.'

The great danger in using television as a visual aid is that it can end up competing with the speaker for the audience's attention. The screen has a powerful pulling force on the human eye, and unless the messages are carefully co-ordinated the audience may be reading the screen and thinking their own thoughts rather than listening to what the speaker has to say.

DIY VIDEOS

The computing companies, most notably Apple with its Macintosh product range, are hoping to do to the video industry what they have done to printing with desktop publishing. They are now producing machines upon which anyone can make their own videos. At present, that basically means putting slides onto videos with some graphics and then adding a voice-over.

This sort of product is of limited use at the top end of the conference and presentation market, since you end up losing both the quality of the slides and the movement advantages of video. But for less formal meetings, where you want the convenience of a visual aid which you can virtually carry in your pocket, it has considerable potential.

Presumably it will only be a matter of time before the systems become more sophisticated and able to handle moving images with more ease. They will then be able to take all the advantages of video presentation to far wider audiences at far smaller cost.

The professionals in the presentation industry, needless to say, are highly sceptical of the sort of results which amateurs can achieve on these machines. It is unlikely, however, that the average managing director is ever going to bother to learn how to use them. Most of the usage will come from the creative departments of the companies, or from their service organisations.

What these machines will do, however, is bring down the costs of producing video back-up material, and help in the overall educational job in the marketplace. People will grow used to seeing them and to being able to use them themselves. Once they have grown used to them in one format, they will look for new and unusual ways to develop them, and so the market will move forward once more.

Chapter 8

Video Conferencing

The global marketplace is now a reality, although not many companies have fully accepted what that will mean to their communications strategies.

As competition for business becomes fiercer, and companies have to spread their resources wider in order to compete effectively, the need for fast and effective means of communication becomes crucial. Television provides an answer to many of the problems.

Just as the telephone changed everything by allowing people to talk to each other virtually instantly anywhere in the world, just as the fax added the dimension of being able to send documents almost instantly to almost anywhere and computer networks allowed for the instantaneous transfer of data, so video conferencing allows people to talk face-to-face and show one another things, via electronic methods. It is one rung further up the technological evolutionary ladder.

It will soon be unthinkable for people to climb onto aeroplanes and spend nights in expensive hotels simply in order to attend meetings which could have been handled on the telephone, had it not been for the need for some visual material, or the attendance of an expert.

THE INCREASING NEED FOR SPEED

The development of jet travel meant that business people had greater access to one another's markets and expertise than ever before. But if a decision has to be made involving people in several companies, you all have to find slots in your diaries when you are free to meet. By the time the meeting has taken place, it might be too late and the competition may have stolen the initiative, either because their diaries were less crowded, they were leaner and hungrier, or they have better methods of communication.

The satellite systems can and do change all the traditional patterns of communicating over long distances. There is no longer any reason why you can't talk to people over the television screens from a studio, or even from the informal atmosphere of your own office.

Imagine that you are a manufacturer with a supplier in Germany. You are designing a new product and several people from your company need to talk to several people from the suppliers in order to discuss specifications and to demonstrate the finer points of the designs. The actual meeting needs to take only a few hours, but if you fly your people from England to Germany, or vice versa, it is going to end up taking at least one whole working day, and possibly two. If you just had to go to a video conferencing studio, either in your company or in the local town centre, you would save a great deal in both money and time.

Video conferencing is as obvious a development of telephones and television as faxing is of telephones and copying. It may seem an age ago, but it is only a few years since fax machines landed on the market. The first people to get them were the companies which could afford to be on the 'leading edge', and the rest of us went on making excuses as to why they were not as useful as they obviously were.

The only limit to their usefulness was the number of people who had them – like telephones they were no good until there was someone else with a machine to take the call. In the meantime we became accustomed to the idea, overcame our techno-fear by using public facilities, and waited for the prices to fall. Now we all wonder how we ever managed without the things.

With video conferencing, there is still a certain awkwardness and self-consciousness when people who are unaccustomed to the medium first talk to the screens. Presumably people behaved in the same way when they first spoke to other people down a telephone at the beginning of the century. We will not find the cameras and screens awe-inspiring once they are commonplace.

At the time of writing, video conferencing in Britain has reached the stage of development which faxes were at in the early Eighties. Most potential customers have heard about it, most can see that it might be useful, but few believe that they can afford to use it at the moment. Things are about to change.

THE PIONEERS

A great deal of awareness about video conferencing in Europe came from US-initiated activity. American companies, used to doing it within their own country, were able to see the immediate advantages of being able to extend their communication networks to their European affiliates and subsidiaries. Slowly the UK business community saw the value of the idea and wanted to be able to take initiatives.

The first people to offer public facilities were British Telecom International and Maxwell Satellite Communications. The Maxwell Group started by running a facility at the Intercontinental Hotel on London's Hyde Park Corner. They retained a suite of rooms which they kept permanently kitted out for transmitting. Up to fourteen people could use the facility at any one time, holding meetings with people anywhere in the world with access to a similar facility. British Telecom, at the time of writing, have eleven such sites dotted around the country, and both organisations can arrange for a studio to be equipped in the city the client is trying to talk to.

It is a costly exercise – about £1,500 an hour to talk to New York, £2,000 to Tokyo and £300 to Manchester. But the video conference organisers justify the expense by comparing their services with flying three people to New York for a meeting. Travelling business class, this would probably cost the client close to £5,000 in fares, plus hotels, food and time away from desks. Suddenly, say the organisers, £1,500 does not look so bad.

NOT AN EXCUSE FOR IDLENESS

It is possible, of course, that video conferencing could be used to get out of doing tedious leg work which is actually essential to the success of a deal. It should not be a tool for the idle.

The public relations group Valin Pollen used the Maxwell suite frequently to talk to their subsidiary Carters in New York when they were taking them over. Without the facility they would have had to fly back and forth across the Atlantic every week. They needed to be in contact constantly, but had to stay in their own countries to service their national operations. All they had to do was tell the organisers the time they wanted to hold the meeting and then they only had to turn up, virtually just down the road, to do it.

The fact that the takeover ended in disaster for Valin Pollen, when the chief executive of their new subsidiary was found guilty of defrauding his clients of substantial sums of money, may or may not have had a bearing on the way they conducted their meetings. Critics of video conferencing might claim that had a little more time been spent getting to know the other players in person, different decisions might have been reached, with very different results for the fate of the group.

When there is a lot of visual material to be exchanged, the video conference comes into its own.

The Advertising Industry

Advertising agency Young and Rubicam, for instance, recently used video conferencing for a pre-production meeting on a product which was going global. London, Washington, Tokyo and Paris were linked, demonstrating the graphics, props and copy. A casting video of the models and 35mm slides were also shown. Other companies like Wellcome in the pharmaceutical sector, and the financial industry, are also good video conferencing customers.

Pharmaceuticals

If a pharmaceutical company is researching a new drug, it will probably have the competition hard on its heels.

It will need to exchange a lot of information as quickly as possible. Some of it could be done over the telephone, with the help of fax machines for transferring documents, but with video you can actually hold the document up to the camera and point to the words you are talking about, or draw diagrams in front of the camera.

The Motor Industry

At Ford they actually wheel-in engine blocks and zoom in on the components they want to discuss, sometimes cutting their lead times by as much as three months on the launch of a new model as a result.

TALKING TO BIGGER AUDIENCES

Once you have set up the infrastructure for video conferencing, it can be used to stage large events more akin to corporate conferences and product launches than informal internal meetings.

Bright Star is an independent satellite operator linking Europe and North America. It is owned by Reuters. The company started in America and was often asked to link with Europe on behalf of clients who found the unliberalised European telecommunication networks too complex to deal with themselves.

The company built up good relationships with the PTTs (post offices and telecommunications authorities) in all the major European countries and it now helps companies with consulting services as well as providing satellite services such as ad hoc video conferencing. When the Eurodollar was launched on the London and Chicago exchanges, Bright Star linked the two locations live, so that people could watch what was happening in real time. They were also able to have a panel of journalists at each end putting questions to analysts over the air.

When one of Glaxo's drugs suddenly obtained a major world market share, the company wanted to let the London and New York Stock Exchanges know about it simultaneously, as the news was likely to cause a major hike in their share prices. The chairman made a video broadcast via Bright Star and was able to field questions from invited audiences at each end.

'Satellite television allows people to talk across different time zones,' explains Frances Whitehead of Bright Star, 'although often people use it for its gimmick value. Live television makes a good focal point for an event, it creates excitement.'

EVANGELISM LEADS THE WAY

Satellite Management International is a company which was formed after staging five days of live satellite-relayed transmissions from Sheffield to sites in the UK, Ireland and Europe for the Billy Graham Evangelical Association in 1985.

During 1987 SMI organised the European segment of the third global Artificial Intelligence Satellite Symposium sponsored by Texas Instruments. They arranged the UK production, an audio-visual link between London and Dallas for a live question and answer session (a world-wide panel discussion to over 1,000 down-links), and transmitted the entire three-hour programme to numerous European venues.

Over 80 per cent of the programme which the audience saw was originated in London, mixed in with video tapes produced in Dallas. SMI are now designing, installing and managing Texas Instruments' new European satellite network.

GASTROLINK

By the end of the Eighties SMI was able to boast of having held the largest educational video conference in the UK, reaching 10,000 doctors with a symposium on gastroenterology.

Gastrolink, as it was called, was staged for Merck Sharp and Dohme, the pharmaceutical company. It featured UK specialists, a surgical operation and a question and answer session between 105 UK sites and another panel of specialists in New York. An audience of 160 was gathered in a London hotel during the day to listen to a series of eminent speakers present their papers. Those presentations were recorded so that highlights could be edited for the evening's two-hour transmission to invited doctors and surgeons at 105 local venues across the country.

As well as the paper highlights, the transmission also covered an endoscopic operation being carried out at the Royal Free Hospital and described live over a two-way audio-video link, and a telephoned question time.

'There was no other way the client could reach an audience of that calibre and size,' says SMI's managing director Peter Thompson, 'allowing GPs access to a dozen acknowledged experts on the subject and not just involving everyone, but answering their questions, either live or after the event.'

The feedback worked via a chain of chairmen. Each local event phoned in questions to the London hotel and they were passed on to the main Gastrolink chairman and programme presenter, who was a professional from Radio 4's *Medicine Now* programme.

WHERE TO START

Although hotels are being used as effective demonstration venues, the future lies with companies having their own in-house facilities, eventually in the form of desktop systems linked to personal computers.

In a hotel, clients talking to Tokyo at the crack of dawn can stagger back to their rooms and get up again in the afternoon to talk to New York. British Telecom provides more humble accommodation for its demonstration centres in its own offices. Clients who use BT's Oxford Street service in London, for instance, are shown how to use the equipment and are then left to confer in private. If they are still unsure of the technology, and do not mind someone else overhearing the meeting, a supervisor is provided.

The organisers prefer to have forty-eight hours' notice, but can accommodate emergency meetings when called upon. The greatest problem is international time differences – it might be easy to arrange a meeting at noon in London, more difficult to arrange it for midnight in Melbourne.

The BT Oxford Street suite is rather like a small boardroom, with a coffin-shaped table and a large television unit consisting of two large and two small screens. The small screens are monitors showing the clients what they look like to the other end, and the large screens broadcast the recipient of the meeting. So you are talking to someone directly on the screen in front of you, just like a television chat show host talking to a guest via satellite.

The chairman of the meeting has a hand control, very like the sort of thing we are all used to pointing at our domestic television sets, only with fewer controls and a list of simple instructions. With this control he or she can move from one person to another, move in for a close-up or out for a group shot, or move the camera on to a flip chart or demonstration model.

There is none of the pressure or tension of a video recording studio – after a few seconds it feels just like talking to the other person face to face.

At the time of writing, BT and Maxwell have reached similar stages in their marketing plans. Both are attempting to entice people to try the product, and to appreciate its uses. Both are also aware that new technological developments are about to open the market up to a far wider audience and bring prices down, perhaps by as much as 50 per cent. The equipment will also become smaller, probably little more than the size of a standard television and video set.

It is not just companies with international links that are taking to the idea. Many of the most successful installations are in UK companies which need to keep in touch with sister offices dotted around the country. Ford and the Royal Bank of Scotland, for instance, both have their own sites set up with BT equipment.

An increasing number of people are taking private lines which can carry video conferences, but they are using half the time for data traffic to make it more cost-effective. Each of these lines is the equivalent of over thirty ordinary telephone lines, and they work on a similar principle to telephone 'hot lines'. As your time limit is booked ahead when you are planning a conference, the knowledge that you are going to be cut off also introduces an element of discipline into a meeting.

Both Maxwell and BT are selling the necessary hardware, and they will soon be joined by a host of competitors once certain industry standards have been established. To set up private sites currently costs about £80,000, and both companies are willing to talk about rental and leasing agreements with clients.

CROSSING OVER INTO THE PRIVATE NETWORK

The divisions between video conferencing, conference broadcasts and private network broadcasting are very slim. The differences seem to come down to the number of locations being linked, the degrees of interactivity and the regularity of the broadcasts.

The strong cross-over between video conferencing and private networks can be useful for a product launch or for a message from management. Advertising agency McCann-Erickson, for example, used it to introduce their new chairman to the rest of the UK company in seven branches around the country.

In another instance MetroVideo, the television and broadcast supply company, was asked by a major insurance company to arrange a sales conference for 1,800 staff by video, using the business television technique. Whereas in the past they had used a five- or six-day roadshow, which took up the time of the top management team and involved a lot of expense, they decided they would stage a show at Olympia Conference Centre. This would then be broadcast live to a number of other sites with an audio feedback, so that people could actually stand up and ask the chairman questions.

The company also linked by satellite to New York and Melbourne to listen to one speaker from each, and the audience was able to talk to them live. Staging the whole thing, including some very expensive video modules, cost around £80,000, which would compare very favourably with the cost of a traditional roadshow, while saving the senior managers a lot of time.

During the Pepsi advertising campaigns, the same technique was used for Michael Jackson's press conferences – allowing the world press to ask him questions live.

The gradual creation of the single market in Europe is bound to accelerate the setting up of networks between countries. Germany is further on than many countries in this respect, possibly because of the number of major business centres within the country, spread over wide geographical areas, but the UK is bound to follow, spurred on by its many links with America.

JOB INTERVIEWS BY SCREEN

One of the uses which the Bright Star satellite is put to is the sending of film rushes and screen tests from Europe to Hollywood. Taking that a stage further, some clients are already using video conferencing for interviewing job candidates across the Atlantic.

Bright Star itself interviews potential American employees from its London headquarters via the television screen. For many people the idea of going to a job interview is daunting enough, and the prospect of doing it on live television would ensure they didn't emerge from their beds for a week. For jobs like the ones Bright Star was filling, however, where confident, out-going people able to think on their feet were needed, it can be a useful way of saving time and travel expenses, and of seeing how potential employees react to pressure.

NOT JUST A MANAGEMENT TOY

The point of video conferencing is that it is not just a tool, or toy, for senior management. It allows other people to attend important international meetings who would not normally be spared to travel around the world. It is much easier to call in support staff to answer tricky questions if they are just down the corridor, not back home on the other side of the Atlantic.

Video conferencing is not a flash in the pan. It is not new technology any more, so it is just a question of time before it becomes commonplace.

PERSONAL VIDEO COMMUNICATORS

For many years science fiction writers have predicted the coming of the video telephone. It has been a long time arriving in reality, but through video conferencing it has now arrived and will become widespread during the Nineties.

A company called Datawize launched a product in 1990 called Universal Video Communicator. If you have a telephone line, they can come to your office and plug in a demonstration model, allowing you to talk directly, face-to-face, to someone on the other end with a similar machine. The system uses a conventional personal computer screen and can capture, display and transmit any amount of motion pictures, and save single frame or full motion video on a hard disk. In America and Japan, large companies like Toyota are already experimenting with personal video communicators internally and looking at ways to market them to the world.

Suppose, therefore, that your employers decided to install one on every desk in your company. You would be able to talk direct to anyone you chose, just as you can on the telephone, but you would also be able to show them things, introduce them to someone in your office, or ask them to show you something. You could be talking to someone in the same building, the same country or the other side of the world. Instead of sending documents by fax, you can hold them up to the camera while you are talking; when you want to illustrate what you are saying with a quick diagram, you just show it to them; when you want someone to okay a design you just . . . and so it goes on.

It brings the telephone call a little closer to being a face-to-face meeting. Suddenly the ordinary audio telephone seems as out of date as the radio became with the introduction of television.

'If there is a video communication set on an oil rig,' explains Mark Webster, manging director of Datawize, 'and another one onshore, it will no longer be necessary for a helicopter to be summoned to take an engineer out to look at a faulty piece of pipe. All they have to do is point a portable camera at the problem and play the pictures into the system.'

Datawize also sells portable satellite systems, which fit into large suitcases and can be set up anywhere.

'When we tell companies what our equipment can do, they don't believe us,' says Webster. 'So we demonstrate it to them, and they still can't believe it.'

WHAT WILL HAPPEN NEXT?

Imagine for a moment that a major company decided to go seriously into this market – let's, for the sake of argument, assume that it will be IBM which leads the way.

Once they have decided that visual communications is the way things are going, they already have the technology in place and the people to install a system which puts a screen on every single desk within the organisation. Wherever there is a telephone in IBM, there could be a video communication screen as well. The company can also use the system for broadcasting private network material, anything from training programmes to employee magazine material.

If the screens are there, then suppliers of IBM could use them as a medium to talk to existing and potential customers within the company, and to help them with after-sales service, but only if they have similar equipment installed at their end. Once IBM has installed a system for internal reasons the chances are that they will by then have a system which they can sell to these people, and will have been able to bring the costs down by using it themselves. They will also want to be able to talk to their customers and prospects on the equipment, saving themselves as much travelling time as possible, and will be selling the system to them.

You then have a domino effect, with every company passing the technology on to their customers and suppliers in an endless chain, until everyone is linked up – just like the telephone system.

FIGHTING THE TIME ZONES

The only limiting factor in video conferencing is time zones. There is little mankind can do to stop the world turning at the rate it does, which is always going to mean that some people are asleep while others are awake – but then the same restrictions apply to long-distance telephone lines, and that hasn't done anything to inhibit the growth in business being done along them.

Chapter 9

Taking Screens into the Public Arena

In an out-of-town Tesco store, a demonstration area is topped off with eight television screens showing customers new ways to use products.

In a two-floor shop in London's Regent Street, a huge screen, operated by laser disk, runs a continuous film of classic scenes from Walt Disney movies with sing-along soundtracks. The latest Disney video releases flicker on three smaller screens, while three-dimensional animated sets put favourite characters through their paces. The Disney studios are as skilful at merchandising as they are at film-making – some might say more so – and they know a thing or two about using television as a point-of-sale tool . . .

Most of this book is concerned with business-to-business applications of television, because that is the area where I believe there is most potential for change at the moment. The public televisual arena is already well exploited in many ways by the broadcasting systems, but there are some highly imaginative ways in which television can be used to bring messages to the public, besides the traditional broadcasting channels.

Changes in technology are affecting the ways that companies talk to the general public as well as to their employees, suppliers, customers and other business audiences. Television screens are, after all, a way of speaking direct to people in as many places as you like at the same time, an ideal way to approach the consumer marketplace.

GOING INTO THE HOME

The obvious example of how business has harnessed this power is with the traditional forms of television advertising. It would be nice to be able to go into everyone's home personally to explain and demonstrate the benefits of your product, but it is obviously not possible, so you have to do it through the medium of their screen.

By choosing that method of communication you gain other benefits which you wouldn't have if you were knocking on doors and talking to people face to face. You are able to build brand images and sell dreams in a way which would be hard to do in person, for instance: you are able to increase a potential market to a far greater degree than would be possible without the help of the medium.

Even this most traditional of modern advertising mediums, however, is now under attack from the forces of change. Competition for viewers' attention is increasing, with new sources of programmes multiplying every year.

On top of which, viewers are now able to 'zap' from channel to channel when the advertisements come on, or fast-forward through them on video recordings. The power that the medium had when it was first introduced has been diminished, although advertising agencies are working hard at creating material which is so arresting or entertaining that it makes people want to watch and keep their fingers off the zap buttons, and media buyers are finding ways of spreading their messages effectively through the new outlets.

TAKING TELEVISION SCREENS INTO THE STREETS

The television sets in the home, however, are not the only opportunities which communicators have to reach the general public. Outside the home there are opportunities to use television to provide novelty value and to give the general public access to events and sights they might not otherwise be able to see.

Tickets for seats to see a great star in an opera house are very dear, but if a performance can be simultaneously transmitted onto a giant screen outside the venue, crowds can gather to watch an event from which they would otherwise be excluded.

The crowds who can't get into Wimbledon can watch some of the tennis action live on 225-square-foot screens sited in a park outside the grounds.

People who attend massive pop concerts or sporting events are often disappointed to find that they are so far from the stage their heroes or heroines are the size of distant performing ants. They are discouraged from complaining by the giant video screens which tower around the stages, giving the same close-up coverage which they are accustomed to seeing on their screens at home.

The politicians have reached the same conclusions, and now have big-screen coverage of themselves at their party conferences, instantly conferring power and status onto speakers who would not be able to command audience attention as effectively if left on a podium with a microphone. Business conferences use the same techniques.

At the same time as sending the live pictures of tennis matches or pop concerts, the controllers of the screens can mix-in video images and commercial messages in any way they like, either to enhance the shows for the audiences, or sell something.

Although the business community has understood the value of this technique for their own conferences, there are other ways in which they could be harnessing its power.

Philips, for instance, have created a product called the Vidiwall. In Sheffield it was installed at an exhibition centre to present a preview of a new shopping and leisure mall due to be opened. Another Vidiwall is then going to be installed in the food court of the shopping centre itself, providing both entertainment and advertising to the estimated thirty million shoppers they hope will be passing through each year.

The potential for retailers within the complex to advertise themselves is tremendous, and the versatility of a system like this would mean that it could be used for highly topical, highly tactical retail promotions.

This is an extension of the sort of video walls which have been appearing in certain high street shops in recent years. They are most notable in the music stores, which have access to the wealth of image-building video material being pumped out by the record companies. The same concept works in any shop with a predominantly young audience, like clothes shops and sportswear outlets. Exciting music and swirling images can lift the spirits – and no doubt encourage spending.

At the same time a continual barrage of noise and movement can exact a heavy toll on the nerves of shop staff who are subjected to it all day long, while being expected to make themselves understood by shoppers and to concentrate on their jobs. If kept under contol, however, television does a good job at providing exciting interior decor which can also draw people into shops or public places, and put across informational and selling messages.

WHAT THE FUTURE MIGHT HOLD

In the long run, perhaps we shall see screens outside police stations running *Crimewatch*-type programmes, reconstructing crimes and asking for witnesses to step forward. Health centres might have health educational films running in their waiting areas, and cinema and entertainment complexes could be running giant trailers for the films happening inside their walls. Large employers might run recruitment films in public places around their premises, and tourist authorities could run informational videos about events and sights to be seen in their areas.

In one way it is a natural extension of billboard advertising, bringing still messages to life just as traditional television commericals brought to life the advertising which had previously been confined to the pages of newspapers and magazines. Instead of coming to Piccadilly Circus to see the lights, the crowds could come to see the cinema trailers and commercials for pop concerts.

The obvious danger is of noise pollution and an annoying clash of visual stimulations in places where people would rather be left alone. But public pressure and the natural caution of planning authorities would almost certainly keep the proliferation of screens under control.

This is more a question of using television on a cinematic scale, drawing crowds rather than communicating with individuals – although the two things are by no means mutually exclusive.

LINKING WITH SPONSORSHIP

With any sort of television which draws large crowds, you have the opportunity for major sponsored events. Futura, the communications consultancy owned by the Trilion Group, has been packaging 'cultural events' which clients can be involved in from the start, and where television can play a major part.

They have one product, for instance, called *Castles in the Air*, which involves taking a tour of classical concerts around the historic castles of England, and tying them in to open-air events like pageants and mock battles, giant laser and firework displays and big screen projection of the concerts to the crowds outside. The theory is that they can create visual extravanganzas for the whole family, which have a high profile in their areas, and then provide merchandising value afterwards in the form of albums, videos and sales promotion materials. Off-the-shelf packages like this, or tailor-made events, can even become permanent exhibitions.

TELEVISION AS CUSTOMER ENTERTAINMENT

In some businesses television acts as part of the service which is being offered to customers: the racing programmes which are beamed into bookmakers' premises, for instance, or the movies and business channels which find their way into hotel rooms and airport waiting areas.

Just as it is good customer-care practice to offer refreshment facilities under certain circumstances – a coffee at the hairdressers, perhaps – it can also be a sales technique to offer specialised television programmes. A business-class hotel which can find new ways of packaging and offering useful business television programmes to customers is going to be able to steal a competitive advantage, if only for a short time, from competitors in the lucrative business travel market. A sports club or pub which can beam-in live boxing matches or other big sporting events, and offer them to customers in some sort of attractive catering package, will have a new product to sell.

Many organisations which have clear ideas of who their client base are and what their needs are, can put together television programming which will be attractive and will act as an incentive to purchase.

POINT-OF-SALE TELEVISION

Television can also be useful as a more direct one-to-one promotional medium. It is already being used in stores as a back-up to the live presentations and demonstrations. Typically you will see a live demonstration of a kitchen blender or a revolutionary new scarf clip, and then the presenter will switch on the monitor to allow a taped version to be run a few times.

There are advantages and disadvantages. Good live presenters will always to able to put more life and sparkle into their performances than their screen versions. They cannot, however be expected to do it endlessly, from morning to night, and still keep up their standards. A video can do exactly that, maintaining a consistency of message.

Some people are also embarrassed to stop and listen to a speaker, particularly if there is no one else listening at the time, but will be less self-conscious about pausing to watch a screen.

INCREASING THE HUMAN PRODUCTIVITY LEVELS

If the promotion is linked to an actual sale – like a scarf clip or blender – then there will have to be someone there as well to convert the initial interest into a sale, but they will be able to talk on a personal level, answering questions and explaining extra points, while the video is still demonstrating the product to newcomers. If it is likely to be a busy stand, this can mean a considerable increase in productivity.

In many cases, point-of-sale material is not expected to make an immediate sale. It could be a promotional film about the wonderful cruise you will win if you collect a certain number of cereal packet tops, or it could simply be a replaying of broadcast commercials.

PROVIDING EXCITEMENT

In some retailing environments, of course, it is impossible to have people stopping to watch screens because they will simply block the aisles and slow everyone down.

The major retailing chains are very fussy about what sort of printed promotions they allow to clutter up the design of their aisles and shelves, and are likely to be even more choosy about a television programme which could interrupt and annoy just as many of their customers as it informs or entertains.

But there are a great many shops where they are lucky to have two or three customers at a time, and clever use of television could help to increase customer loyalty and to demonstrate goods.

Imagine, for instance, a boutique in a suburban town, supplied by moderately expensive designer labels. If there was a television in the corner of the shop, showing videos of the clothes on sale being worn in a fashion-show setting, or being photographed on models, customers would be able to have a much better idea of how the designers envisaged the clothes being worn than can ever be achieved on a clothes hanger or in a fitting room mirror.

A DIY store could be showing videos in different sections which explain how to use certain kits, and demonstrating just how good they will look once assembled.

An estate agent could have a series of videos of houses on the books, either playing constantly to attract people in, or to be selected and shown to customers when they make specific enquiries about houses in specific price ranges.

A garage providing a service for a car – or even just a tyre change – should be able to sit the customer in a lounge area where programmes are playing which demonstrate new car models, new tyres, and possibly advertise the services of the AA and other similar organisations at the same time.

INTERACTIVE POINT-OF-SALE

Point-of-sale is another area which lends itself to interactive material, if the customers are sufficiently interested and confident to use it.

If for instance, a motorist was having problems with his or her car, and wanted to know what to do about it, an interactive video provided by the manufacturer of the car and on display in a local spare parts shop, could lead the motorist to be able to identify the problem and purchase the right part for the job.

TOUCH-SENSITIVE WINDOWS

The use of interactive videos and disks can even provide information for window shoppers after closing time. A product called Telesensor provides touch pads which are attached to shop windows. The customer, passing by the shop at any time, can touch numbers on the pad and bring information and videos up onto the screen of a television in the window display. The sound is transmitted through the glass by special loudspeakers.

The idea has already been put to use by banks, building societies, travel agencies, computer shops, insurance companies, post offices, public transport offices and tourist information centres. It means that window displays in town centres can be working twenty-four hours a day, providing information and product demonstrations for which the customers are actually asking.

THE FUTURE

As the general public becomes more accustomed to two-way television, the screens are going to provide companies with another way of reaching out to talk to their customers.

They will be able to undertake research, by having interactive sets asking questions of passers-by in the same way that human researchers do now. They will be able to demonstrate new products, and they will be able to provide expert assistance at all times of the day – even when the shops are shut.

Chapter 10
Television as a Selling Aid

Using a screen to help make a sale is one step further on from using it as an advertising or promotional aid at point-of-sale. Rather than waiting for the customers to come to you, however, you could actually be taking the television out to them.

THE PROBLEM OF PORTABILITY

The inhibiting factor until now has been the size of video players. While they are technically 'portable', they are still the weight of an average sewing machine, and are not therefore something that every travelling sales person is going to want to cart around on calls.

There are going to be exceptions to this, where a potential sale is of sufficiently high value, or where the use of moving pictures transforms the sales message, but on the whole it is impractical to expect sales people to carry the players.

Using the Customer's Own Machine

The alternative, which becomes increasingly viable, is using the customer's machine. In most offices it is now possible for someone at management level to be able to gain access to a television and recorder, if they want to see the film badly enough. If you are selling to people in their homes – whether double glazing or pension plans – the chances are again high that there is a machine available.

The problem for the sales person to overcome is the feeling of taking liberties with the customer's property at a time when the relationship may still be very tentative.

If the situation is strong, however, and the customer has expressed an interest in the product, then there should be no problem. A video can then be used to demonstrate a number of different things. It could show the size and reliability of the organisation with which the customer will be dealing. It could be used to show how the product works, and take the customer to demonstration sites which might take hours or days to visit in the normal course of events.

If, for instance, you are selling an engineering product to oil companies and you want to show that it can survive freezing conditions in the Arctic as well as the heat of the desert sun, nothing will do that more effectively than film of it actually being used in both locations, together with some vox pop recommendations from satisfied customers.

Anything involving a demonstration which would take time and money to arrange live will lend itself to a video promotion.

VIDEOS AS DIRECT MAIL ITEMS

If we assume that most potential customers will have access to a video machine, we can then begin to consider the possibility of sending them a cassette through the post, either as a cold-call item – although this would be enormously expensive with current technology – or as a follow-up to an enquiry for a brochure or some other promotional item.

A customer might, for instance, send off for a brochure on holidays in the Caribbean, or for information on having a swimming pool built in the back garden. The brochures would come back as promised, and in them would be an invitation to view a video. In the case of the holiday company, it would be looking at the various properties on offer, their views of the beach, the dining-room, and some of the sights around the island. From the swimming pool company, it would be film of a pool actually being built and then some atmospheric shots of finished pools of various shapes and sizes being used by happy families. In both cases there might also be vox pops of satisfied customers.

Naturally this is an expensive form of promotion, but the costs could be helped by charging a nominal sum for the video, either refundable on return of the cassette, or deductible from the price of the product if a purchase is made.

By the time potential customers have got as far as thinking about a high-cost purchase like a holiday or swimming pool, and by the time they have received the company's literature and have confidence that they are respectable, they will probably not resent sending off £10 in order to find out more, particularly if that £10 is refundable.

The actual mathematics of this sort of operation would depend upon the quantity of tapes being sent out, the price of the product being promoted and the likelihood of conversion to a sale. If the sums can be got right, however, the principles are sound. Nothing sells a product as well as a successful, practical demonstration. Television can do that perfectly. It can also create atmosphere and show new perspectives on old subjects.

As database marketing becomes more and more accurate in its abilities to target audiences with precision, the possibilities for sending out reasonably highly priced pieces of television to potential audiences become ever greater.

A hairdressing salon with a good list of potential local customers could send a video of their people in action, creating different styles for models. A garden centre could send out film of shrubs and trees that they have in stock, with demonstrations of how to plant them and film of mature gardens to tempt the customer to invest. A car dealership could send out videos of a new model in action, full of atmosphere and recommendations from the right sorts of people. The programme could include television commercial material, test-drive footage with a famous racing driver, and technical engineering information on why the car is so exciting to drive.

The principles are all the same as catalogues and other direct marketing items, and it is already happening in many companies.

INSTRUCTION VIDEOS

Another area where print has reigned supreme until now has been in instruction manuals, but there is no reason why some of them could not be converted to television with a consequent improvement in clarity of message and image.

If you have just bought a new car, and you are feeling particularly proud of your choice, you will happily sit down for half an hour to watch a video telling you how to maintain it and run it successfully. The manufacturer or dealer who supplies the video could use it as an opportunity to build not only your loyalty to the brand but the sales on other services, too.

In the business-to-business arena, instruction videos become closer to training films. If your company has just bought an expensive piece of computer software, a well-made video explaining to the users within the company how to use it and what benefits it will bring, will put a friendlier face on the technology, helping to overcome any techno-fear which might be lurking within the company and speed up the learning process.

In the case of a complicated product like software, it might even be worth creating some interactive training material.

MINIMISING THE WORK LOAD OF THE SALES FORCE

As the cost of keeping sales people on the road continues to grow year by year, every company needs to look at ways in which they can keep their costs to the minimum while still winning new business.

Just as the telephone has come into its own with telemarketing, and the computer has led to the growth of database marketing, television programmes can add a whole new dimension to face-to-face selling.

Compared with print, the making of videos is going to be expensive for many years to come. Compared with the cost of a live sales call from an expensive member of a sales team, however, a video which sells seems very good value indeed.

VIDEOS AS PERSONAL PRESENTATION AIDS

Everyone is looking for new ways to help them put across their points at meetings. It started with chalk on blackboards and moved on to flipcharts and overhead projection.

Although it is commonplace to see videos used as speaker back-up in conferences and major presentations, most people still consider it too expensive a medium for everyday use.

In many cases they are right, unless you work for a company or in an industry which uses film and video regularly, and therefore has large libraries of archive material which can be quickly edited to illustrate a point.

'Video is very flexible,' says Maxwell Michie of Shell. 'It can be shown today and gone tomorrow. We can very quickly assemble continuum loops for exhibitions, for instance, or small programmes for staff meetings, which cost as little as £1,000 to put together. That is another advantage of an in-house facility. It would be hard for an outside production house to do anything for those sorts of prices.

'The result of using video widely is that it improves the general standards of communication around the company. After it has been done once, the demand tends to grow. If your colleague has just given an important presentation including video, you aren't going to want to revert to chalk-and-talk again.'

Chapter 11

Education and Training

Imagine for a moment that television had been invented by the Ancient Greeks. Imagine how it would be if the British Museum had archive footage of great teachers down the centuries talking to their students, or speaking direct to camera. The potential list is endless; Plato, Aristotle, Leonardo da Vinci, Charles Darwin, Bertrand Russell . . . we could take interior design courses with Michelangelo, writing lessons from Charles Dickens, drama workshops with William Shakespeare and poetry readings with Lord Byron.

(It is interesting to speculate whether these great figures of history would have become victims of overexposure, showing up human weaknesses and failings which history would otherwise have been able to ignore, undermining their own potential myths.)

Future generations, however, are going to have just this gift of first-hand experience of the greatest teachers of our time.

We are now able to encapsulate the endeavours of our brightest and best, and distribute them to audiences who would never otherwise have had a chance to see them. We should, in time, be filming all our professors and gurus, thinkers and writers, artists and philosophers at their peak, and maintaining libraries of videos just as we keep libraries of their writings.

The downside to that, of course, is that it is impossible to predict who will prove to be the greatest in the long term, and so we will also have to keep an awful lot of mediocre material as well – just as we have with the written word.

Broadcast television has been doing just this, but only up to a point. When a great statesman, writer or thinker dies, we are instantly given access to filmed records of them in action at various stages of their lives in the form of obituary programmes or 'tributes', as they are more frequently called.

VIDEO IN HIGHER EDUCATION

At the beginning of the Eighties, when the business community was still unable to see any real use for television, the educational establishment was quick to grasp the value of the medium. Many of the universities, polytechnics and educational authorities set up video units and started to explore ways in which they could make the screen work for them.

'We have a video unit,' says Colin Burgess, head of Media Services at Brunel University, 'which does a mixture of promotional work for the university and educational programming.'

The promotional material is designed to sell the university to potential students by giving them a feel of how the place works, while the educational programmes are generally made very simply in the university studio – rather like filmed lectures, but with added inserts if demonstrations are needed.

'The educational programmes are particularly useful for distance learning,' says Burgess. 'The problem tends to be that most academics find it hard to give up the time needed to make a programme. So they tend to do what is convenient for them. If, for instance, they are called upon to do a particular demonstration over and over again, they will put it on video, so that it can then be put into the library for people to look up when they are attending the relevant lecture. Many of the lecturers are good presenters and come across well on the screen.

'Some of the laboratory experiments and demonstations have very long lifespans. The only things which date are the fashions, and presenters with long hair and wide ties tend to raise a laugh, but the theories remain the same.'

The university also uses the facilities to run video courses as part of their communications and information studies, and for the learning of presentation skills.

'In a perfect world,' says Burgess, 'we would have programmes on all the lectures in the library, just as we have printed material. We should also be ablc to market some of the products of our better-known lecturers to the outside world. We do get asked to produce programmes for outside sources, like a consortium of companies called Europace, which includes names like IBM and Hewlett Packard, who use us to make computer science programmes based on simple lectures. They then send them out on satellite to their offices around the world.'

EDUCATION IN THE BUSINESS MARKETPLACE

Now the business community has been the one to catch on to the idea, although the products are designed to appeal to audiences today rather than to posterity. The aim in most cases is to find new avenues down which business thinkers and trainers can spread their teachings more widely and effectively, and sometimes to make the exercises profitable as well.

Electronic performances will never replace the well thought out book, but they can bring elements of books alive and make them accessible to a wider public. It is one thing to read the lengthy books of management guru Tom Peters, quite another to see the man himself in action.

Similarly, television talking heads will never be quite as satisfactory as seeing someone live or debating with them, just as watching a concert on television does not hold the same thrill as buying a ticket and actually being there.

The question is, would you have bought that ticket anyway? Television does not replace the live performance it is filming, it merely takes it to a larger audience, to people who would never normally think of going to a concert.

When you take that theory and apply it to the spreading of a training message through a large company, you can see what the advantages of television might be.

How long will it take a team of trainers to go through a company of twenty or thirty thousand people? It could take a year or more. With the help of television and carefully planned training packages, however, you can retrain an entire workforce virtually simultaneously. As the pace of change and competition increases, companies need to be able to react quickly to new initiatives and ideas, and television gives them the means to do that effectively.

THE INTERNATIONAL GURUS

James Martin is one of the most successful business and high technology gurus in the world. In 1969, while working at IBM, he published a book which forecast the rise of technology in the Seventies and Eighties with remarkable accuracy. In it he claimed that electronic technology would enable the best teachers to earn more than the best film stars.

Within ten years, Martin had a net worth far in excess of almost any film star, and a lifestyle to match. His fortune has been built from his writing, teachings and predictions about the future. One of the major outlets for his work is a library of video programmes in which he talks to camera about developments in the fields of technology. The videos, which are also computer based, are then rented out to companies, universities and colleges to form video libraries. He has made over 240 programmes so far.

The Smithsonian in Washington is one of the world's grandest and largest museums. In its Museum of American History they have created a video wall out of dozens of television screens to form the climax to a permanent exhibition on computers. The wall shows many different images simultaneously, but, occasionally they link together to show one large face, one of several experts talking about the future, Martin among them. The film had to be processed in Japan because only they had the computer technology necessary to create the many synchronised images.

To buy a seat at one of Martin's live seminars would cost several thousand pounds, and would involve travelling to one of the world's capital cities at a time when his road show was passing through town. But a video, once rented or purchased, is available constantly, and can be controlled to suit the student's speed of comprehension. Some might like to watch it straight through and then run back to vital parts; others might like to stop and make notes at various stages, or simply stop to think about what has been said. All that can be done at the touch of a button.

Always looking for new and better ways to use the media, Martin became the first person in the world to make an educational product on laser disk, and then the first interactive educational laser disk. Then came the possibility of linking computers to the laser disks and of creating hypermedia, for which he had the software created. This led him to exploring the possibilities of building artificial intelligence into documents and creating an electronic version of a book with computer pictures, sounds chunks and television included.

Martin explains, 'Marshall McCluhan said, "Thc medium is the message". This is an exaggeration. What McCluhan meant is that the medium *changes* the message. You can say different things in a book from what you can say on television. Lectures can be more powerful than video, but video can say some things that live lectures cannot. Hyperdocuments enable their user to navigate through vast bodies of information if they wish, and they can have expert systems built into them. Multimedia presentations can be very powerful if the human presenter remains in control.

'Too often today the medium swamps the message. Television is dominated by television professionals, the cinema is dominated by cinematographers who will spend $60m on a movie with a trivial script but incredible visual craftsmanship. For valuable transfer of knowledge, the master of the subject must dominate and the medium must be a tool for communication.'

There are a number of big international business stars whose videos are distributed everywhere. Many of them are best selling authors as well, like Tom Peters who was launched to stardom when he co-wrote *In Search of Excellence*, and Kenneth Blanchard who first hit the big time when he wrote *The One-Minute Manager*. Both these stars are distributed in Britain by a company called Melrose. Michael Porter and other stars of the Harvard Business School are distributed by Video Arts.

In Britain the first television star of general management training is Sir John Harvey-Jones. With an impeccable record as a businessman, having been a successful chairman of ICI, he started to become known to the general public through broadcast television programmes such as *Question Time*. He then sold the idea of a series to the BBC, with himself as the star, called *Troubleshooter*, in which he went into small and medium-sized companies as a company doctor, diagnosing and making recommendations.

The mixture of real-life drama and Sir John's robust personality caught the imagination of a large audience, and he is now marketing his own video products on other subjects, all of which he originates and writes himself.

TAILOR-MADE OR GENERIC TRAINING PRODUCTS

The first question facing a company which has identified a need for television training programmes is whether to commission something in-house, or to buy something 'off the shelf' from one of the generic training companies. These generic products are designed for, and marketed at, the corporate training department. An individual is unlikely to be able to afford them.

There are a number of firms offering these packages, the best known being Video Arts (famous for its John Cleese films); Melrose Training; Playback (founded by comedians Mel Smith and Griff Rhys Jones); the BBC; Longman, which has bought the old Rank training films business; and Gower.

Wyvern Crest, the business book club whose advertisements appear in many of the national papers, and whose direct mail shots arrive on the desks of virtually every small and medium-sized business at regular intervals, has a sister company producing corporate training videos. The videos, however, unlike the books, are aimed at the corporate rather than the personal purchasers, teaching 'soft' skills like how to question people and how to handle constructive criticism. The company is run by ex-publisher Hugh Murray.

'Although I would dearly like to create a mass market for training videos,' says Murray, 'I keep finding myself forced back into the corporate area. It seems that the figures just don't add up when you start selling videos at £10 or £20, because there are a limited number of people who buy self-improvement products.'

In fact, probably the same people who buy the books also buy the videos, but the books cost half as much to produce, which makes the sums very different. In the long term it seems certain that there will be a personal market for business videos, but at the time of writing it is hard to see where the breakthrough will come from.

There must be a middle ground somewhere between the Tom Peters and James Martin products in the business sector and something like a Jane Fonda exercise tape which has achieved huge distribution to the personal market all over the world, just like a feature film. Until that happens, however, the generic training films will still be sold to large companies rather than to individuals, where a price tag of £1,000 or more is not a problem to the customers. Although there is some cross-over between the generic and custom-made products, they are very different industries.

The firms with generic products are marketing and selling them in just the same way as a music or book publisher. Like other businesses, they need capital to set up and they need to research and test their products thoroughly if they are to be successful. Those making commissioned programmes are in the same boat as general production companies – in fact they usually *are* general production companies.

Throughout the Eighties training companies like Video Arts and Melrose have grown steadily and profitably, seeming to be much more solid and structured businesses than the companies selling production services rather than specific products.

'Before a company even considers whether they are going to buy or commission,' says Peter Cole of Melrose, 'it must be very sure about what it needs to do, what the objectives are and what the resources are.'

There are some areas which will obviously be particular to one company, such as product or procedural training, whether it is training sales people in the benefits of a new product or bank clerks in how to speed up their services to customers. These programmes will obviously always have to be tailor-made.

Marks and Spencer, for instance, makes all its own training material, apart from management training films which it buys from the major companies. Each of their stores has a training co-ordinator who has a library of the material which has been made in the past, and will design programmes specifically for the needs of that store, using the programmes as back-up. Each of the programmes has about a three-year shelf life, which means there are about thirty current ones at any one time.

The programmes are prepared with outside production companies, and are sent out with guides on how to use them and support material to check if people have actually learnt anything from them.

SPECIALISED AREAS

There are some highly specialised production companies creating programmes in different areas (such as the financial world), who may have generic products in the area that you need, but if not they will have the background knowledge to understand your problems quickly and suggest solutions which are based on experience. If, however, a client company is clear enough in the objectives of the programme, any professional production house could make training material in the same way as they make other types of corporate programmes.

Products for the Financial Industry

Financial i is a company, now owned by Euromoney Publications, which produces video-based information packages designed specifically for the international financial community.

The films are full of computer graphics and worked examples, and present complex material in a concise and step-by-step way. The sort of subjects they cover are *An Introduction To Foreign Exchange*, *Currency Options* and *The Eurobond Market*. They have forty different titles in all.

'We are covering esoteric material,' explains Colin Sullivan, director of marketing for the company. 'We are assuming that the viewer knows nothing, but has an intellectual ability. We market to the human resource and in-house training departments of banks, accountancy firms, corporate treasury departments, financial directors and their staffs, government agencies, software houses and law firms. In most cases the trainers who use our products will not know much about the subjects we are covering, so we are providing them with tools to help put the messages across.

'I believe that there is no other medium which can impart as much information in as little time as television. It provides viewers with a sense of "being there", which they can never get from the printed or spoken word alone. We can explain something to them with graphics and then actually take them down onto a trading floor in London, Tokyo, Hong Kong or anywhere else to see how the product is actually traded.'

Sponsorship Opportunities

Making the films, which involves a lot of work on location as well as the computer graphics costs, is expensive. To help them offset the costs, Financial i seek sponsors for their various programmes. Price Waterhouse's name, for instance, appears on one of the packages, and Lloyds Bank on another. A sponsorship package costs in the region of £40,000, which is only a percentage of the costs involved in creating one of the programmes.

'As well as the videos, we give customers work books,' says Sullivan, 'so that they have the material in print as well if they need to look back or check on something. We also provide questions and answers to ascertain their levels of comprehension, and there are interactive floppy disks for the trainers.

'Video should never be a stand-alone product. It is just one more tool which a trainer should have in his or her armoury, and it is an effective way of reinforcing points and illustrating different aspects of the subject.

'Television helps to bring otherwise dry subjects alive by breaking them up into digestible segments and by varying the pace. It wakes audiences up. It is also possible for the trainers to freeze a television picture at any stage to demonstrate a point, using the screen as an electronic flip chart or blackboard.'

TAILOR-MADE PROGRAMMES

The obvious advantage of tailor-making programmes is that they can be made personal to the company. Characters can be seen actually walking in and out of your offices, and real people can make guest appearances.

The most obvious disadvantage is the difference in cost, which is enormous. Not only does it cost more money, it also takes a long time to start a production from scratch, and uses up a great deal of management time. There is also a risk involved, because you are setting out to create something which is unknown and unseen, against a generic product which has already been tested in the market and can be inspected and evaluated before purchase.

Video Arts also makes commissioned programmes for clients, using their star-studded cast of television comedy writers to personalise the product to the client company. The initial research on the project is done by training specialists and the writers are then given highly detailed briefs on what the client company is wanting to achieve.

All of these writers have experience of writing shows for broadcast television and are therefore used to working within half-hour time-frames. The results should consequently bring humour to bear to subjects which most people are sick of hearing about (safety, for example – a favourite topic for training films), while ensuring that the message is always relevant to that specific client.

The danger with commissioned material is that someone within a client company may have personal ambitions to be involved in the making of a television programme and they may commission when they should buy a product off the shelf which will cost 5 per cent as much, or even less.

In some cases video or television might not actually be the best medium for the job at all.

It is not, for instance, very good for conveying a great deal of detail to audiences who already have a strong fundamental understanding of a subject and are wanting to move on at their own speed into more esoteric realms.

In those sorts of cases a linear training video could actually slow things down (we will talk about the interactive alternatives in chapter 13). So any training programme requiring attention to a mass of details would be better to stick to print and the spoken word. It is, however, highly effective at conveying the softer subjects, like interpersonal skills and general man-management techniques.

Getting the Brief Right

If you decide to commission a new programme the briefing document which you give to the production company will be critical. Just as with corporate videos, the clients must be clear who they are talking to, and what they want those people to learn or do differently.

Once the brief is completed it would be a mistake to go out to too many production houses with it. Make a short list by talking to people with experience of the industry, and then commission the company of your choice to do an outline of how they would approach the project.

It is important to consider how the audiences are going to see a video. Will they be watching it formally in a group training scheme? Will it be part of a large-scale conference or event? Will they see it running in the reception area of the office? Will they be asked to take it home and watch it with their families over the weekend?

Once they have seen the programme, what activities will follow? Will they be able to talk to someone about it? Will they need to write an assessment of it or make a report on how they can use what they have learnt to improve their work performance?

Other Materials

Training manuals can demonstrate how the video is just part of the overall training scheme, and give instructors ideas on how they can fit it in with their own training schemes, or how to run a complete course with it.

The Melrose generic products, for instance, come with such manuals, whilst some of their products build up into two- or three-week training courses, with the videos as just one element.

'In most cases our customers are experienced trainers,' says Melrose's Peter Cole, 'so they usually look through the books and find ideas which they can incorporate into their own schemes. That is the best way. If you just stick people down in front of a video and expect them to learn something in those ten or fifteen minutes, you won't achieve any results.'

There is a place, of course, for the well-made training programme as entertainment, slotted into a heavy training course in order to change the pace and give some light relief. It is important, however, that the trainers doing this do not expect to get any results other than an improvement of mood.

TESTING THE GENERIC PRODUCTS

If you are coming into the generic training video market for the first time, you will need to find out what is available. In large client companies there will already be a wealth of information available, perhaps even a library of existing titles which have already been bought or leased.

Companies like Melrose hold preview days in venues around the country to allow companies to view new products. They also work through preview centres in all the major cities where potential clients can go in and view material which they think, from catalogue descriptions, might be useful to them. The preview centres work on commission from the production companies. You could also go into the companies themselves, most of whom have viewing rooms for just this purpose.

Other methods are trailer tapes, which work as omnibuses of products just like previews for feature films. If it is a major product, like one of Melrose's two-week training courses, the trailer tape will have detailed explanations of how the products work and sample sequences.

A number of the companies also operate short-term hire schemes, so that you can get the product for a day or two in order to assess it and work out how you could build it into your training programme. Often client companies start by hiring programmes and trying them out, and then end up buying them outright, once they have decided that they work for them.

Chapter 12

Training in Communications Skills

There are two reasons why you might want to use television cameras to improve your communications skills.

The first is the obvious one of needing to learn how to appear on television in order to overcome nerves and be competent when asked to appear on broadcast television or in a corporate programme.

The second is to use the camera as a way of seeing your non-televisual performances objectively. The latter might be needed to rehearse a new business pitch, or simply to see how you come across in a role-playing situation or while delivering a speech.

In either case the camera is just a tool to allow you to make objective assessments of your own style, and for trainers to demonstrate to you your communications weaknesses and show you ways to improve them.

There a lot of people offering training courses in television presentation skills. Many of them are people who have made a name for themselves in broadcast television and are looking for ways of broadening their earning potential.

Most of the courses serve some purpose (there are some which are complete confidence tricks to part punters from their money, but they are fairly easy for a businesslike buyer to spot). Amongst the bona fide training companies, some are more effective than others. How effective they are for you will depend on your needs and the professional abilities of the individual trainer.

WHO NEEDS IT?

'I think a lot of companies buy more management training in this area than they are ever likely to use,' says Roddy Gye of GHA. 'It is a bit like buying a Swiss army knife with twenty different gadgets for survival in the wild, when all you are likely to need to do is cut bits of string.

'Anyone who is going to have to talk direct to camera, however, will have to be very fluent, and will need to be trained. No amateurs should be allowed to attempt it.'

WHAT WILL THEY TEACH YOU?

Professional trainers can teach you how to speak convincingly to camera, and they can teach you some of the ground rules on how to be interviewed. They will teach you how to deal with a hostile media during a crisis, for instance, or how to render stilted speech in a fluent and gripping style. How successful they are will largely depend upon the personal skills of the teacher and the relationship established with the pupil or pupils.

Sophisticated television hardware will always help to demonstrate points and smooth a training session along, but it is not as important as the calibre of the people involved.

'We tend to bring in well-known interviewers and journalists from the broadcast companies when we want to tune up our executives,' says Maxwell Michie at Shell.

'These people are expensive, and we could do it ourselves, but using someone well known helps to concentrate people's minds. What happens is that I act as the producer, explaining what television organisations want, and then the journalist does two interviews. The first is a gentle, laid-back programme which allows them to talk as much as they like. Then we do a hard-nosed interview in which they are asked awkward questions.'

Alan Watson, a television journalist who started his training career with the BBC, training business people to appear on *The Money Programme*, is now a director of Corporate Vision Ltd. in London, which provides high level training as well as production services from their newly equipped basement studio.

In the years between these two jobs Watson was a reporter for *Panorama*, a radio broadcaster for the *Today* programme and he made a number of televison documentaries. He left the BBC to become head of television and radio services at the European Commission and then moved into communications consultancy.

'There are a number of ways in which closed circuit television can help to enhance executive communication skills,' he explains. 'It could be a straightforward need to prepare for a radio or television interview. It might be to rehearse a speech for a conference or annual general meeting, or a preparation for making a corporate video. It might be for a company to rehearse competitive pitches, or television might be used as a presentation aid as part of a general management training course.'

REHEARSING COMPETITIVE PITCHES

With increased competitive pressures in all commercial sectors, companies are being forced to mount new-business pitches to win orders and projects where previously they might have expected to merely pick up repeat business. With contracts sometimes worth many millions of pounds, it is essential that companies pitch as well as they possibly can.

Executives who might once have relied almost completely on documents to do their selling have now realised that to succeed they need to be exceptionally skilled at spoken communication and all the other aspects of a face-to-face presentation, where potential clients are making judgements about a company based on the people they are meeting from that company.

Television training is one way for executives to find out what sort of impact they are making, to improve it and to assist in getting across key messages.

'The Japanese are particularly aware of the importance of presentation,' says Watson. 'One of our first clients was a Japanese bank which was competing for the job of handling the privatisation of a British company on the Japanese stockmarket.

'A team from Tokyo came over to London and spent three days in our studio, rehearsing everything from the handing over of business cards at the beginning of the meeting to the actual presentation itself. They were tremendously self-critical of everything they saw played back on the screen, and they built a great *esprit de corps* during those days, which also helped them in putting across the message.

'They were completely committed to winning, and they didn't intend to take any chances. They wanted to know exactly what they looked like as they came into the room, and what they sounded like when they made their pitch. They wanted to know how to work with the audio-visual material they planned to use; how to present as a team, with the right degree of interaction between the main team members; and how to handle question-and-answer sessions.

'They wanted to understand fully what preparation was needed, the relationship of the actual presentation to the accompanying written material and, above all, they wanted to identify and then deliver the Unique Selling Point of the presentation, so that it was entirely persuasive.'

The same sort of dedication to presenting a professional image, and using television to hone it, now spreads through every industry where big-money contracts are at stake. It could be insurance or advertising, accountancy or manufacturing. Television allows people to see themselves as others see them, and to correct flaws which might otherwise have gone unnoticed.

'It is particularly important for the top people to watch themselves on the screen,' says Watson, 'because if they are bad presenters no one within their own companies may have the nerve to tell them. But if they see themselves doing it badly they can do something about it.'

The ability to communicate effectively to key internal and external audiences has become an essential management skill. The successful executive has to convince and persuade a variety of different people in a variety of different situations.

Employees, shareholders, customers, legislators and the media all have the power to enhance or damage a business. Managers need to be able to communicate information effectively and efficiently, persuasively and attractively. Closed circuit television can work as an 'electronic mirror' in which executives can see, evaluate and improve their performance with the help of experts.

CRISIS MANAGEMENT

Crisis management consultancy is a growing business. As the media become ever broader and ever hungrier for scandals and disasters, companies have to be prepared to be pounced on when something goes wrong, and to have their mistakes broadcast across the world. If they are not skilled in handling the media, the stories will become exaggerated, elaborated and worsened.

Someone who conveys the wrong image and messages could become a villain in the eyes of the beholder; the same person handling the problem in a different way, might emerge a hero.

The problems might be financial disasters or environmental ones, plant closures or industrial accidents. They can happen to any company at any time. In some industries, like oil or chemicals, where the risks of disaster are high, companies tend to ensure that all their senior managers are thoroughly trained at all times for the unexpected occurrence.

For most companies, it is probably not worth going to those lengths, although some basic grounding in the necessary skills will always be useful to managers anyway. When a crisis does arise, the company needs to know who it can turn to for fast advice. That skill will probably be inside their public relations consultancy, who will have the necessary communications experts on tap – to work through the night if necessary. It could equally be a television training company.

COMMUNICATING A TOTAL IMPRESSION

Television has a number of characteristics which pose problems as well as providing opportunities to interviewees. Being a visual rather than a verbal medium, it is impressionistic. A printed transcript of the actual words you use will not tell you much about the effectiveness of the total communication.

You may have been saying positive, optimistic things, but did you look positive and optimistic when you were saying them? Your appearance and your manner should enhance what you actually say.

Television is a very intimate medium and the slightest change of facial expression or demeanour will instantly register with the audience. Never forget that they are judging you as a personality at the same time as reacting to the message you are trying to put across.

WHO ARE YOU TALKING TO?

Appearing on broadcast television will mean that you are talking simultaneously with audiences whom you would normally address separately and in differing styles. You are perhaps being watched by your superiors, your peers and your subordinates, your shareholders, customers and suppliers.

The audience profile might include members of parliament and housewives, senior civil servants and school leavers, potential customers and competitors. You can't hope to please all of them, but you can find a style which hits the middle mark more often than not.

Although broadcast television audiences are measured in millions, most of the viewers are watching in small groups in their homes. So a rhetorical style of presentation is absolutely inappropriate.

TIPS FOR APPEARING ON SCREEN

Broadcast television usually offers no chance of control, or any opportunity for preparing what you want to say in the same way as you can prepare a press statement or a written publication. You can, however, prepare yourself in other ways. By being well trained in the whole subject of personal communication, you will be able to handle unexpected situations with greater skill.

It is always better to be prepared for any situation, and a television interview requires as much preparation as possible.

If the interview is to be pre-recorded it will help if you know who else they are interviewing, how long the item is to be, when it is going to be broadcast and what angle they are taking on the subject. The chances are that you won't be given an answer to the last question, but if you have all the previous information you will be able to make an educated guess.

If the interview is to be broadcast live, then you need to know how long it will be, who else will be in the studio, and whether the discussions or interviews will be preceded by film. Again, you want to know their angle, if they have one.

When a firework manufacturer was asked by a television company if he had any footage of beautiful firework displays, and was invited along to chat about the subject, he was delighted by the idea and did not ask any questions about the style of the programme. Only when it actually went on air, and pictures of children burned and mutilated by fireworks came up on screen, did he realise why they had brought him in, and by then it was too late to escape.

Although live broadcasts are more frightening, you do at least have more control over the final product, provided you know what you are in for and have prepared in your mind precisely what message you want to get across. If it is pre-recorded, you have no idea what is going to happen at the editing stage.

If you are going to be edited, don't talk for too long. If they are looking for a two-minute piece, don't talk for thirty minutes. Ten will be enough, otherwise they will have too much material to work with and may lose the sense of your key message.

You can emphasise your key points, and thus show the editors the best bits to include, by headlining them as you speak with 'This is the main point', or 'For me the most important thing is . . .'.

Once you know what they are after you can prepare for the broadcast offensively, not defensively. Decide what it is that you wish to communicate before worrying about what you don't want to discuss. Get your message clear in your mind and be certain that you can express it simply and clearly without resorting to jargon. If you need the support of any particular statistic or quotation, write that down and have it with you. Do not try to memorise lines or write down whole speeches in advance – you must come across as if you are talking spontaneously and conversationally.

If the cameras have come to your premises, be careful about your surroundings during the interview. Take a conscious decision about what you wear and about the decor of your room or office. What are you trying to communicate? Authority? Intimacy? Remember that viewers will be looking at what is happening around you as well as at you. If they see bookcases full of intelligent books, they will make a judgement about what sort of person your surroundings suggest you are; if they see a swimming pool full of sun-bronzed young people, they will make a different judgement.

If you are doing a live interview in a studio, make sure you get there in good time, and don't have more than one drink. Never confess to anyone that there are areas which you do not wish to discuss or they are bound to bring them up. Allow the make-up artists to help you: they are professionals and they know what they are doing.

Make sure that in the studio you can see a monitor and hear the sound track of any film which precedes the discussion, so that you can see the context into which they are putting you.

'No one who is new to television should try to learn a script,' says presenter Roy Sheppard. 'If it helps their nerves they can prepare some bullet points to remind themselves of the key points they want to put across, but no more.

'Sometimes with a corporate video, executives will be given entire scripts a few days before, which they will sweat blood to learn, only to turn up on the day of the shoot and find that the words have been changed.

'They also don't need to worry about mistakes made during recordings, because they can always be edited out and covered by camera shots of the interviewer nodding wisely. It is much better to keep going in a conversational way than to keep breaking off and asking to start again.

'It is vital that they resist the temptation to look into the camera the moment the sequence is over and ask, "Was that all right?". It is a temptation which few can resist initially, because they are so relieved to have got through the experience, but it ruins the whole shot. It doesn't leave the editors any room to play with when they come to do the cutting. You should always leave two or three seconds before saying anything or, to be on the safe side, you should wait until someone else breaks the silence.

'There are certain rules about clothes as well. Black and white is bad, and red bleeds into everything around it on screen. Pastel coloured shirts are better than white, and small checks and patterns are bad because the 625 lines on the television can't handle them and the result is an uncomfortable strobing effect. It is also not a good idea to wear jewellery which is likely to jangle while you are talking, or strapless dresses, because in a head and shoulders shot you will look naked.

'On the whole the interviewees are in more control of the situation than they think. Usually the whole thing is over so quickly they can't believe it has actually happened. A three-minute interview is a long time in television terms, but flies by for the participants. The important thing is not to try to make too many points too quickly. About one point per minute is the maximum, regardless of the questions being put.

'Interviewees tend to see themselves as victims, but they aren't. They are actually being offered very good opportunities to say what they want, and once television producers have found that someone can deliver the goods, they will call on that person again and again.'

ALWAYS BE YOURSELF

Lee Bowman is managing director of the Kingstree Group, a management consultancy team dedicated to helping senior executives (not to mention the odd European crowned head and American president) to improve their methods of spoken communication. Part of their brief sometimes includes preparing people for television interviews.

'Being directly questioned about anything, even in a friendly fashion, is slightly unnatural,' he explains. 'A professional interviewer does not have time to go through the polite rituals of getting to know you, as someone would in a social situation, before starting to ask you personal or difficult questions.

'The result is that most inexperienced interviewees become nervous and act out of character. This is a disaster, making them seem insincere and uninteresting.'

When clients go to Kingstree they are sat in front of the camera without warning and asked some general questions, like what their companies do and what their jobs involve. These are questions to which they know the answers all too well, and that is where they go wrong.

'We can show them the results on tape immediately,' says Bowman, 'and they are able to see how important it is to be prepared. Because they have given no thought to what they are going to say, what audience they are addressing, how long they are going to be talking for, or what their key messages should be – because we have given them no warning – they look and sound terrible.

'This acts as a warning to them of just what is going to happen if they do not prepare. If they can't talk effectively about their jobs and their companies in a relatively informal atmosphere, how are they going to cope with any difficult questions which might come along in a more formal setting?'

PREPARE YOUR KEY MESSAGES

The illusion that many people have when going into an interview situation, according to Bowman, is that because they know more about the subject than the interviewer (which will always be the case because that is why they have been invited), they will be able to 'busk it' and talk off the top of their heads. This is a grave mistake.

To give the best possible performance, it is not possible to be thinking up what you are going to say whilst worrying about how you are going to say it.

'It is critical that they work out what their key messages are and how they are going to put them across before they walk out in front of the cameras,' explains Bowman. 'Then their minds will be free to deal with the stresses of adrenalin and to work out the best way of delivering the messages.'

According to the Kingstree philosophy, there should be no more than three key points to be made, and there could be less. Each one needs to be illustrated with examples or anecdotes to bring it alive in the minds of the listeners.

'The more you know about a subject,' says Bowman, 'the more crucial it is to prepare in advance, because of the time limitations. The average time for a television interview is two and a half minutes, which gives you very little opportunity to put across a lifetime's experience of a subject. If you have too much knowledge of a subject you will be tempted to go into too much detail, which will result in a loss of the audience's attention and a failure to put across the key messages.'

It is true that having intelligent things to say doesn't necessarily make a good television performer. Many of the politicians, showbusiness personalities and trade union leaders who are most frequently quoted in the media are the ones with the least to say. What they do have to say, however, is very clear and concise, and frequently repeated.

An interviewee can also take control of a conversation if they are clear about their central points. They can start by stating their case, and that will lead the interviewer to adjust the questions which follow, giving the interviewee the initiative.

'They should avoid using any sort of jargon,' says Bowman, 'and they should treat the interview exactly as if it were a conversation with an acquaintance who has a great interest in, but no technical knowledge of, what they are saying.'

There is no fixed formula for success in dealing with television cameras. What is essential, however, is that there is a real commitment on your part to communicate, and that you take the time and trouble to prepare. Do not underestimate the power which television has to help your cause, and also to hinder it.

Chapter 13

Interactive Video

As technology marches onwards there are going to be more and more fusions and cross-fertilisations amongst the different disciplines. The two most obvious bedfellows for corporate television are telecommunications, (as seen in video conferencing and 'business television' broadcasts), and computing.

The merger of video techniques with computing started happening during the Eighties in the world of interactive video.

Interactive video means the merging of video and computing technologies in order to add the realism and immediacy of the television picture to the technical versatility of computer based and computer assisted training. The ability to put live, full motion, full screen video on your personal computer screen is one of the most exciting and potentially revolutionary developments in communications technology.

Overlay video boards, which combine live video with computer generated graphics, provide users with the potential for developing a host of applications, from new video production techniques and training videos to corporate video conferencing. They allow users to connect their computers to a video source and play live videos on their screens, offering them the potential for combining video from an outside source with computer generated material such as sound and animation.

TRAINING USES

So far, training has been the largest user of these types of technology.

Financial, retailing, management and technical skills have all been the focus of major training programme developments by organisations like Allied Dunbar, American Express, Austin Rover, Lloyds Bank, Price Waterhouse, B&Q, Littlewoods, British Telecom, British Steel, Shell International and Jaguar. There have also been major point-of-sale initiatives from people like Abbey National, Ford and Pira Unipart.

The concept works at every level, from programmes teaching spot welders how to do their jobs, to programmes from Price Waterhouse, the accountancy giant, teaching their people how to interpret financial statements.

At either end the programmes can be made participational and entertaining. (The Price Waterhouse product, called *Account Ability*, was structured so that viewers learnt about the subject while investigating a suspicious death in a dramatised parody of the *film noir* detective thrillers of the 1940s.) After using the programme at their own training centres, Price Waterhouse distributed it commercially to clients like the Bank of England, Bank of Scotland, Bankers Trust Co and Citicorp.

At the other end of the market it is possible to make training programmes for people with very low levels of reading skills, because the questions can be asked by a voice-over or presenter.

GIVING THE VIEWER CONTROL OF THE MATERIAL

Interactive video involves the making of programmes which the viewer can control, move about and influence at the touch of a few buttons, or even at a touch of the screen.

In most learning situations a teacher drives the proceedings, and an audience is largely passive. A traditional video training programme, like any other sort of television production, is a linear affair. The producer of the programme decides what you, the audience, will see and you then have to start at the beginning and watch through to the end. You have no control, it is a one-way communication.

Travel at Your Own Speed

If the programme is interactive, however, it will ask the user questions at certain stages. Do you understand that so far? . . . Do you want to know more about this, or that? The user then presses the buttons or touches the screen in response and the computer moves on or goes back accordingly to the appropriate stage.

As a result no time is wasted going over stages which the users already understand, and at the same time they are not moved forward until they have completely mastered what has been taught so far. The users are able to pose questions, ponder problems, explore the subject at their own speed and receive answers.

WHO COULD USE IT?

The uses to which this can be put are diverse, and new ones are coming up every day. Allied Dunbar, for instance, uses it for sales training, Austin Rover uses it to provide sales people with product knowledge and health and safety information; they then use the same material for demonstrations at motor shows and in their showrooms. American Express uses interactive video to support their sales presentations at board level in large companies.

From cockpit simulations for trainee pilots to babycare advice for new mothers in hospital, interactive video can fulfil any number of functions.

An estate agency, for instance, could put together a database of 45,000 different properties, with any frame able to be recalled within seconds. A customer could come into the shop, select a type of property, an area or a price range, and the computer would put up on the screen a list of likely properties available. The customer would be able to choose some and the screen would then show videos of the inside and outside of the properties, and the customer would be able to make print-outs of the ones they wanted to see.

Once the general public is more confident about using them unattended, customer kiosks could spring up in a number of locations. Anywhere where we might currently see sales demonstrations, for example, such as department stores and supermarkets, could be an opportunity for interactive video machines. They are already appearing at special events like exhibitions.

British Airways used interactive video as part of their *Day in the Life* programmme of staff motivation. They would start with a theatrical show for audiences of twenty at a time. A short video would then be followed by a competitive quiz, where the audience was divided into teams.

The Ministry of Defence uses it for training people to drive Chieftain tanks, or to play games purpose-designed to promote a greater awareness of Saudi Arabia among fighter pilots, using existing film footage linked to computer graphics.

The Midland Bank had customer information video machines designed into their branches, with the hardware usable for staff training as well.

Shell has developed a touch screen/laser information system of maps for motorists, containing route information and warnings of traffic problems.

3M made a video to go in their reception area on which visitors could select any division of the company and see a range of the products available.

By far the largest use of interactive video is in the training fields, but it is also used extensively for marketing projects like point-of-sale material, trade shows and exhibition stands. It is also useful in education and for the storing of reference material in museums and similar institutions.

HOW IT IS DONE

To construct an interactive video, a television programme is broken into segments by the makers, and these pieces can be manipulated to provide an array of different paths for the viewer to follow through a presentation.

It is a blend of audio, visual and textual databases and computing technologies. The images of still or moving pictures with two separate audio tracks are held on videodisc, while the text and graphics are held on the computer. The videodisc recording is made up of many individual frames, all numbered and indexed like the pages of a book. Through these numbers any segment or frame of the video can be retrieved at random, and a series of still or moving pictures can be presented in any order.

In other words it provides 'random access'. The body of information can be followed by questions, responses, answers, decisions and choices from the viewer. Although the computer holds a massive amount of information on the given subject, it presents on the screen only what is relevant to that individual.

Traditional teaching methods have always included the use of audio-visual aids, but they are still only able to make learning a passive experience for most people; a typical audience for a trainer or teacher is *receiving* information rather than *experiencing* it. Interactive learning, however, brings actual experience within the grasp of the viewer. It creates a dialogue and immediate feedback between the teacher and learner, and provides real life, hands-on experience.

Someone learning how to weld a car engine actually has to decide where the welds should go; a trainee accountant actually has to demonstrate an understanding of the subject in order to move on to the next stage; a trainee pilot will actually 'crash' the flight simulator into the skyscrapers of Hong Kong, if he or she makes the wrong decision while landing at the island's airport.

HOW DO WE LEARN?

Research by the British Audio Visual Society highlights the importance of being able to teach through a variety of media. According to them, we remember about

10 per cent of what we read
20 per cent of what we hear
30 per cent of what we see
50 per cent of what we see and hear
80 per cent of what we say
90 per cent of what we say and do at the same time.

In the US IBM found that an interactive training video for new managers increased the delegates' mastery of the course material by approximately 300 per cent over conventional classroom methods. In a point-of-sale context, the Co-operative Wholesale Society found that an interactive programme resulted in a perceived increase in sales (of products promoted by the programme) of between 15 and 20 per cent in a controlled test. A programme produced by Asda increased sales of specific products by 45 per cent.

With video it is possible to convey real life experience, thereby motivating and involving the audience, and conveying a consistently clear and cogent message and image. You can demonstrate large and complex products reliably and to their best advantage, and you can provide consistently accurate and current product information, such as prices and availability.

The Abbey National, for instance, provided a point-of-sale video containing information on a range of products and services. The software which they used allowed mortgage and interest rates to be continuously updated, and it also recorded how often the services had been selected by customers.

With some systems you could even complete a sale at the time and place of the product demonstration or promotion, by building an ordering system into the programme. It just requires one more question to the viewer – 'Do you want to buy?' If they press 'Yes', the computer can then take their instructions and process them.

WHY VIDEO AND COMPUTING MAKE GOOD BEDFELLOWS

Video is primarily a visual medium, with audio, text and graphics working to support and clarify the messages. It is best used for presenting dynamic processes but less useful for static content. It is good if you need motion, or if you need to present actual events to the audience. It works for dramatisation and historial re-enactment, or for compressing and expanding time. It is effective for micro and macro photography.

Video is also able to provide a fixed standard of presentation – in other words, television and computer programmes do not have 'off days' or get tired at the end of a long day. They always supply teaching at the optimum standards which were achievable at the time of making the programme. The medium is highly effective at simulating hazardous events in a safe environment, with flight simulators being the most dramatic example.

Computer technology, on the other hand, is primarily a text and crude graphics medium. It is best for the presentation of static content and modest animation, with only a minimal facility for audio at the moment. It is, however, usually highly interactive, with the programme producer able to lay a multitude of paths of travel through well thought out instructions, giving the viewer the ability to control the order of presentations.

Computers are good at handling content which requires learner activity, practice and participation, with the learner able to control the pace and sequence of instruction. It is a flexible form of presentation based on individual needs.

Computers can provide drill and practice of basic concepts, and reinforcement based on learner responses, plus the ability to keep score and to provide control to the learner who wants access to various portions of the instruction. They can provide simulations and problems which respond to a learner's control of events. They do, however, require the viewer to have some basic reading skills, which may not be necessary with video.

CUTTING LEARNING TIMES AND COST

Companies which have used interactive video training report some startling results. Jaguar claim, for instance, that it cut learning times by an average of 50 per cent, and always provided the most efficient path to content mastery, while being easily understood and adaptable to different learning styles.

In 1987 they were launching the new Jaguar XJ-6 internationally, which was placing an enormous pressure on the dealer training department. The complexity of the car's system required extensive training for technicians; given the shortage of trainers and resources, this proved logistically and financially impossible. So they commissioned Chrysalis Interactive Services to produce a series of six interactive video programmes.

The videos took training directly to the workplace. Jaguar estimated at the time that it cost them roughly £300 per day to send one person on a conventional training course. In 1987 the interactive videos provided 2,400 training days with the new two-day courses, where conventional training would have taken 4,800 days.

Alliance and Leicester Building Society identified a significant gap in staff knowledge reflected in poor sales, so an interactive video was produced. An independent evaluation of the project showed that performance in the pilot branches improved 77 per cent and account sales went up 40 per cent.

One user in America wrote that by using interactive video he was able to give his people a complete course of training in the time that it used to take them to drive to the location of the traditional training course – which, incidentally, lasted fifteen days.

Users also claim that it reduces training costs, particularly if it is going to be seen by large numbers of people, since the majority of the costs are incurred at the design and production stages, not at the replication, distribution or delivery end. Federal Express, for instance, has been quoted as expecting to save over $100m by using interactive systems for employee training. British Telecom was able to replace a one-day training course (which cost £150 per head) with a video which worked out at £15 per head and achieved better results.

Privacy is a major factor in the effectiveness of this medium. Many people are embarrassed in traditional training groups to admit that they have not understood something, and human teachers can lose patience with slow learners. It is also embarrassing if the whole group has to be held back while something is explained to you again. With a video being shown to you alone, you can go over the material as often as necessary until you have complete mastery and are ready to move on to the next stage. That ensures that all training programmes have strong foundations, and are not based on a false assumption that someone has understood the basics before moving on to the more advanced stages of the subject. 'I can train myself!' says one enthusiastic user. 'I can understand what's going on.'

Equally, someone who has grasped the basics quickly can move on at their own faster pace and doesn't have to wait around while the slowest in the class catches up.

Research amongst users has shown that people retain over 25 per cent more of what they have learnt when it comes to them through an interactive programme. They find that, by working with simulations of the actual equipment or processes which they are going to be using in their jobs, they can raise their confidence and be entirely proficient by the time they are actually put to work.

The individual's motivation to learn is increased by the responsive feedback of the technology, plus the individual involvement which means fewer distractions and a greater focus of attention.

Interactive video is not just a solitary method of learning; it can also be used to good effect in groups. Natural events, for instance, can be shown on the screen and then paused for a moment of decision or controversy which will stimulate discussion and responsive group dynamics. It can provide entertainment, dramatisation and enhanced presentation within a traditional training setting.

AVAILABILITY

Availability is another key selling point. There will always be a shortage of good teachers in the world. If the work of the best ones can be captured in an interactive manner it can be disseminated to as wide an audience as necessary, where teachers, as mere mortals, would be limited by their own abilities to be in one place at one time.

IT'S FUN

More important than anything else, perhaps, is the finding that people actually enjoy working with interactive videos. People who are enjoying themselves will always learn faster and better than people who are bored or frightened of failing. By and large, manuals are boring, videodiscs are fun.

A training system which is well liked by users and achieves high levels of learner success, which also achieves significant cost benefits for applications involving a large requirement for distributed training, and which compresses the time usually required for training, is likely to become commonplace. With the introduction of digital techniques which increase the speed and flexibility of the programmes, and with the application of artificial intelligence which increases the power of the systems' creators, the whole sector is going to be moving up a notch in the evolutionary scale through the next decade.

There have been predictions from the experts that within a few years most personal computers will have sound capabilities to match their visual powers, and within IBM they are working on a machine with accurate voice recognition entailing a vocabulary of several thousand words.

VIRTUAL REALITY

Taking the twin concepts of interactive video as a private medium and as an experience simulator to their ultimate conclusions, the boffins talk about equipment which they refer to as 'virtual reality'.

In the most popular visions this consists of a helmet with internal 3D displays, together with electronic gloves that sense your finger and hand positions. People are also talking of electronic chairs, and creating computer generated worlds full of realistic 3D scenes and stereo sounds that include you within them by sensing your movements.

They might, for instance, be able to give you the entirely realistic experience of being two inches tall and clambering around inside a car engine, or of being a thousand times smaller and taking a guided tour of the insides of a calculator. Equally, on the grand scale, they could take you hang gliding, or Formula One racing, in the same way as they take pilots through the flying procedures of jumbo jets in flight simulators.

Just as the Sony Walkman can make a giant pop or orchestral concert into a personal experience for a listener, Virtual Reality equipment could add vision and touch to the equation – personal television on a breathtaking scale, a cross between a ride on Disney World's Space Mountain and the old pantomime trick of putting a comedian on a tea trolley and running a speeded up film of the M25 on the backcloth.

It all sounds like science fiction at the moment, but then fifty years ago many of the machines we work with today were beyond anyone's imagination.

COMMISSIONING AN INTERACTIVE VIDEO

Commissioning an interactive video is not unlike the process required for the more straightforward programmes, except that the costs may be higher and you will have to deal with computing people as well as film people.

Start with a detailed tender document, outlining your audience, your aims and objectives and the scope and content which you want the programme to have. Tell them your time, budgetary and other restraints, and outline the benefits you hope to achieve. You could also give some initial ideas on implementation strategies and evaluation. You should ask for an estimation of the cost, and then for a presentation outlining how the company would approach the project. This will provide opportunities for clarifying key issues raised in the tendering and proposal documents and will give both sides the chance to decide whether they can establish a good working relationship.

Once you have decided who to work with, the more autonomy and trust you give them the better. Clients who want to be involved with every stage of the development will always slow it down and push up costs.

FOR FURTHER INFORMATION

The National Interactive Video Centre (NIVC) at 24 Stephenson Way, London NW1 (tel. 071–387 2233) was established in 1984, offering objective, impartial advice and information on all aspects of interactive media for training, education, marketing and communications. They offer free demonstrations on the basic principles of interactive technology and run a full range of courses. Private briefings for organisations can also be arranged.

Chapter 14

Getting it Seen

There is no point in making a programme if no one is going to watch it.

We talked earlier about researching the audience in advance and making sure that you know exactly who you are talking to and what they want to hear. Even if you do that you still won't be able to get them to sit down and watch the product unless it is marketed to them in the right way . . . Assuming, of course, that it isn't a compulsory programme, such as part of a training project.

The problem with television is that it is an intrusion into people's time. If you send someone a printed catalogue or brochure, they can flick through the pages to see (*a*) if it is the sort of thing that is of interest to them and (*b*) which bits are relevant and which are not. Having done that, they can then make a decision as to whether to read it there and then, throw it away, or store it to read later.

With a television programme the decisions are different. They have to be able to decide from the accompanying literature whether or not they are likely to be interested in what the cassette or broadcast contains, and once they have made the decision to watch, they know that they will have to sit right through the programme if they want to find the bits which are of interest to them.

And where the brochure took a few seconds or minutes to deal with, the television programme will take longer – perhaps as much as half an hour. If a video is relegated to a shelf to be looked at later, it has probably missed its chance of being seen because other, more urgent, tasks will pile up ahead of it.

PACKAGING

All the onus therefore, rests on the accompanying documentation, which will include the actual packaging of the cassette itself.

If it comes in a plain black box with a typewritten label stuck on the side, it will not have done anything towards tempting someone to open it. Effective packaging might put as much as 10 per cent on the costs of creating a video, but if it makes the difference between it being seen and not being seen, it is a small price to pay. Just as people are used to high standards of broadcast television, they are also used to the high standards of video packaging which they see in their local video store, and they will assume that anything less is a reflection on the company which has made the programme.

INVOLVING THE AUDIENCE LIVE

Another way of reaching a specific target audience is to lure them with the mystique of a television studio. Even in these sophisticated times, it is surprising how many people have still not witnessed the making of a television programme.

When BOC wanted to inform an important part of the investment community about its achievements in the world of medicine, and about its plans for the future, they decided to make a video. To add some glamour to the proceedings, and to involve the target audience in the event, they recorded it at Limehouse Studios, then in London's Docklands. It was a full-blown show, complete with a well-known presenter from the broadcasting world, revolving hospital sets full of actors playing scenes using BOC equipment, and presentations from senior executives.

The event was created by BOC's corporate design consultancy, Pauffley and Co., and was built into their overall marketing campaign. It worked, with remarkably senior analysts coming down from the City for lunch on a boat tethered beside the studios and a glimpse into the world of corporate showbusiness.

As the business community becomes more accustomed to the wonders of television, this sort of event will have less pulling power, but at the moment it is still an effective incentive if staged cleverly enough and if the information being imparted is suitably important.

STARS AND PRESENTERS

Another way to persuade people that your video is worth watching is to decorate it with a star name. This is not always a good idea.

The presenting, interviewing and acting stars of broadcast television can make very good second livings from the world of business television, and there are some strong opinions within the industry about whether or not they merit the enormous fees they can command.

It is important for the client company to get the balance right. If you hire a big star – Terry Wogan, Noel Edmonds or someone equally famous – you are going to gain prestige by association, and you may also be able to increase the number of people who will sit through your programme. These 'stars' do, after all, have an appeal to certain sectors of the public, otherwise they would not be as successful as they are. The danger, however, is that they will eclipse the client and the client's message. People will remember that they saw Terry Wogan or David Jason and that they were very funny, but they may not remember the messages which the programmes were intended to put across.

When it comes to programmes which require acting skills, the same applies: a star name may distract from the message. At the same time, however, the difference in acting standards between well-known and unknown actors can be enormous.

As with so many things, it comes down to basic experience. An experienced television and film actor will have a better chance of turning fairly pedestrian scripts into something interesting; less experienced actors may not have the confidence or expertise to do that, so will be at the mercy of their material and their directors.

If the directors are good enough, that will not matter, but often tight budgets mean that results have to be achieved quickly and time cannot be spent coaxing a performance from an inexperienced performer.

Big names are particularly successful when comedy is involved. Often humour is a good way to get a message across, particularly a training message, and humour handled badly is the worst possible embarrassment. Consequently, stars like John Cleese, Mel Smith and Griff Rhys Jones are able to work highly effectively in the business market.

John Cleese is a particularly good example with his Video Arts training products which have become comedy classics whilst putting across strong messages at the same time. In less skilful hands, the impact would have been greatly diminished.

THE RELIABLE ANCHOR-PEOPLE FOR LIVE BROADCASTS

In the more serious world of presentation and television journalism, ranged just behind the big stars are well-known and trusted faces from the world of broadcasting like Nick Ross, Valerie Singleton, Peter Hobday and William Woodward. They are solid professionals with years of experience of dealing with the problems of live broadcasting.

For the large-scale live shows such as conferences, these people's fees pale into insignificance compared with the overall budgets involved, and they seem to represent good value for the services which they perform.

Their professionalism and experience mean they are able to smooth over the rough bits and give a professional gloss to shows which might otherwise be a series of technical disasters. They often bring a much needed air of calm confidence to proceedings which can be fraught and tense. While everyone else is losing their tempers and panicking, it is often the presenters who supply the sense of perspective and humour necessary to carry the project to success and hold the team together.

Having said that, there are also some prima donnas who do the exact opposite, but they are generally well known within the industry and are not usually at the top of producers' lists when they are considering who to approach. The ones who are used most often are the ones who have proved their worth time and again.

'These people are an enormous help to the executives involved,' says Martin Pentony-Woolwich of Spectrum Communictions. 'They can handle all the links and continuity, and they ask the questions, allowing the executives to concentrate completely on the answers which they give. If we are doing a live broadcast to several locations, using an audio feedback, and one of the locations suddenly goes down, the chief executive of a company is not going to know how to deal with it.

'But a professional presenter will be able to respond to a voice in his earpiece telling him to move on to another location, without the audience being aware of any hiccup.

'Anyone who has dealt with hugely complex technological achievements like the telethons, where there are thousands of things which can go wrong, will be able to bring an air of authority and calm to a business broadcast.

'Professional presenters make the people they are talking to look good. Most of them are journalists by training and, by bringing an enquiring, journalistic manner to the proceedings, they accord them the same professionalism as a broadcast programme.'

Bringing Credibility With Them

These television personalities also bring with them the credibility of the programmes with which they are linked. It might be *Top Gear* or *The Money Programme*, breakfast television or *Panorama*, *Tomorrow's World* or *Sportsnight*, but the personalities will provide an endorsement for the company involved simply by being there, without having to do anything other than explain the corporate messages and ask the questions.

'These people tend to be successful for two reasons,' says Pentony-Woolwich. 'Firstly, they are highly professional in everything they do, so they are hired again and again, whereas anyone who behaves like a prima donna will never be reused. Secondly, they have a presence and a charisma which lift the level of the proceedings.'

There is also a layer of younger and less well-known faces who are highly professional and understand the needs of the business market. In some cases a client might decide that it would actually be detrimental to have a 'known' face, and will go for someone less famous. Such people, however, are hard to find, because the very nature of their profession means that before long they get known through working for the broadcasting stations. But as the number of career opportunities for presenters and journalists increases, so there will be a larger pool of anonymous professionals on which to draw.

Roy Sheppard is a newsreader and presenter from programmes like *Open Air*, who has made a specialisation of business programmes. 'It helps the people being interviewed,' he explains, 'if the interviewer understands their business, at least a little. There are some presenters who simply work from a script, and are unable to ad lib or ask a supplementary question on their own initiative. This results in a very stilted and amateurish dialogue, getting no further than the standard corporate clichés, and will not put the interviewee at ease.

'If the presenter or interviewer is actually understanding the subject, the interview will become a conversation rather than a set piece of propaganda, and the audiences will be able to see the difference.'

It is important that anyone involved in a live broadcast for a client should spend some time getting to know the company in advance.

If they do not fully understand what it is the company is doing, and what the key messages are that it is trying to put across, they might get an inflection wrong which would completely destroy the credibility of themselves, the other participants and the programme itself.

WHY HAVE A PRESENTER IN A VIDEO?

'Being a well-known television face can be both an asset and a weakness,' says Michael Rodd of Blackrod, himself still remembered by many for his work on *Tomorrow's World* in the Seventies. 'For every one person who is attracted by your past reputation, there will be two who are repelled.

'Presenters aren't always a good thing to have in a business programme anyway. They can sometimes get in the way of the message. The time when they are useful is when there is something to do on camera, like drive a truck or a car, or demonstrate a product. They are also useful for explaining things when there aren't any pictures available to do the job. A good journalist should find something out and then yearn to share the knowledge with an audience. The same applies to presenters.

'If you have neither of these situations, however, why do you need a presenter? It all depends what sort of story you want to tell and how the client's evidence needs to be presented.'

The presenter can also be a useful liaison point with the client on the day. 'If someone from the client company turns up on the shoot,' says Sheppard, 'they are often puzzled by what is going on – or more often by what appears not to be going on. The technicians are usually too tied up with what they are doing to take time to explain themselves to the client, but the presenter often has long spells of waiting around, and can take on the job of liaison.'

For corporate videos, where the pace is more leisurely, it is not critical to have someone with a high degree of experience, and the difference between the costs of an unknown professional actor or journalist compared with a 'name' may make the difference between the budget being workable or impossible. A well-known name from the world of broadcasting can charge around ten times as much as an unknown, (perhaps fifty times as much in the case of a star).

Indeed, the days of the big stars are over, Roy Sheppard believes. 'Most business people have ceased to be impressed by all that. The fact that the Radio Rentals Rentathon was fronted so successfully by someone from their own staff shows how things should be done.'

THE DISADVANTAGES OF STARS

'There can be disadvantages to hiring a star to talk for a company,' says Roddy Gye of GHA, 'because everyone watching knows perfectly well that the stars have been hired for the day and that they aren't part of the organisation. If you make a programme with George Cole, for instance, and the script requires him to say, "We at Bloggs Widgets do this, and believe that . . .", it is obviously going to look like a put-up job. An unknown face or voice-over, however, can look like part of the team, even if they have only been hired for the day and given a script to read.'

The other drawback is that a well-known face might date a video badly. Someone who has a very successful broadcast series running at the time you are making your programme may not work on anything so high profile again. Alternatively, they may retire or even die. The viewing public is very fickle, and tends to associate certain performers with certain periods.

If you want your video to last a long time, and in the case of a training film it might be as much as twenty years, it would be a mistake to tie in with a face which will label the product as being out of date.

Chapter 15

Editing and Post-Production

Editing and post-production are the bits of the process which probably interest the clients least.

As long as there are scripts to look at, casting sessions to attend and shoots to go on, the clients know where they are: Hollywood, a familiar territory for anyone brought up in the age of film and television.

Once the programme goes into the hands of the editors, however, it is a world of technicians and technologists. It seems less accessible and more esoteric. It is at this stage, however, that a programme can be saved or ruined; it is also a stage where a budget can be saved or broken.

ENDLESS POSSIBILITIES FOR SPENDING MONEY

The possibilities for spending money in post-production facilities houses are endless, and the more advanced and sophisticated the gadgetry, the wider the scope for unlimited spending.

Much of the equipment might be necessary for producing a big-budget commercial which is going to run several times a night for months on end: if an audience is going to be subjected to the same piece of programme so many times it has to be perfect, or the faults will become glaringly obvious. That doesn't mean to say, however, that every piece of business television with any budget at all shouldn't look entirely professional and up to the standards of broadcast programming.

It is at this stage of the proceedings that a client needs to trust the judgement of the production house. They need to be able to hand over the rough cuts of the film, and feel happy that they are in the hands of professionals who will not run up unnecessary expenses.

The Off-Line Edit

The first stage will be 'off-line editing' – which is something akin to working with proofs in the printing world. At the proof stage you can still make changes without holding up the presses or incurring massive resetting costs.

Off-line edits are done on non-broadcast quality tape, on equipment which is quite cheap to hire but which produces results sufficiently good to allow the producer and the client to make creative judgements. In other words, you will be able to see what the finished product will look like, even though it still looks a bit rough.

The On-Line Edit

Once everyone is happy with the off-line edit, it will go to an on-line suite – which has to be hired by the hour at extortionate rates – and the final job can be done in a matter of a few hours, when it could have taken weeks to get to that stage without off-line editing.

It is in the area of post-production techniques that technology is probably moving the fastest in the film and television industry, with new equipment coming onto the market all the time. Much of it, however, still does not conform to common standards, which increases the confusion and the potential expense. It is an area where the client badly needs a trusted guide.

CLIENT INTERFERENCE

'Clients always like to interfere with the making of their corporate programmes.' This quote, in various permutations, comes from the mouths of so many corporate television makers that it is pointless to attribute it to any of them.

The problem seems to stem from two facts. The first is that everyone has watched television a lot during their lives, and consequently believes that they know how to do it, that it is easy. The second is that an awful lot of the people who commission corporate programming are stage-struck, at least to some degree.

'It is amazing,' says Roddy Gye of GHA, 'how many clients will decide they have to attend the shoot on the day you are taking the helicopter up to get some aerial shots. They pretend that it is because the shot is crucial to the video, but they spend the day sitting there open-mouthed with awe.'

Clients should try very hard not to interfere. If they have chosen the right production company and if they have briefed them fully and correctly and are happy with the treatment or script, they should hardly need to be involved at all until the final product is delivered – apart from making sure that the production team has access to all the people and facilities which it needs for the shooting.

Providing Technical Advice

Sometimes, if the subject involves a lot of technical detail, then it is important for someone from the client side to be there to advise. It is no good making a technical film about servicing an engine if something in the engine is obviously in the wrong place, and it might require the eye of a trained engineer to spot the error.

If there is going to be a representative of the client company on set, however, it must be someone with the authority to make decisions there and then. If the director, editor, writer or presenter wants to know whether they can do or say something, they need to be able to get an instant answer. It is not helpful if the only person there is from the public relations company or department, and has to act as an intermediary with the client. Every hour lost through that sort of behaviour puts up the final costs.

ANIMATION

Cartoons, the sort we are used to seeing from the Disney studios or the broadcast television equivalents, are enormously labour intensive, needing animators and their assistants, tracers and painters to create the thousands of necessary frames. They are consequently highly expensive to produce. Indeed, on the very big-scale productions, where they are going to need literally hundreds of people, the film makers tend to go out to the Far East, where labour is somewhat cheaper.

There can, however, be situations where animation can be useful for providing a change of pace within a corporate video, or for providing a moving logo which is going to last some time and be reused in different contexts.

'Animation can be useful in things like medical films,' says Rob Dunbar, production supervisor at Filmfair Animation, 'showing things like the effect of a pill on a molecule, something which you couldn't do with live action but which can be easily achieved with some cheap animation.'

If you want animators actually to create characters for you, however, to the standards that you see on broadcast television, there are no short cuts.

'We produced a video for schoolchildren thinking about careers,' says Shell's Maxwell Michie, 'with the objective of alerting them to the exciting world of science, and new-materials sciences in particular. The programme was very densely packed with information, and we thought one way to break it up would be to introduce cartoon characters, showing the molecular structure as animated creatures.

'The danger with using cartoons to reach this age group was that we feared we might make it seem too juvenile for them, so it was a difficult creative brief initially. We asked a number of animation studios to give us some rough ideas and in the end the results were very good, but we went through the most enormous amount of anguish along the way, and the process used up a large proportion of the budget for the programme.'

Making the Logo Move

Mike Lodge, one of the partners in Mike Lodge Animation, says, 'What usually happens is that the client gets graphic designers in to do the initial concept work and then they call us in to see how we can animate it. It might be something as simple as a company logo which they want us to animate.

'Computers can do a lot of the work, making shapes and letters tumble around the screen and look like chrome and so on. But you still have to generate each drawing, colour it and pay for all the equipment, which often makes it more expensive than traditional animation techniques. Computers, however, can provide much neater and slicker packages when it comes to things like lettering and pack shots.'

Merging with Computer Images

'Sometimes we are asked to combine our work with computer images. One of the computer companies had created a high-tech shoe which they wanted to make fly, but they couldn't make the computer draw wings which looked real. So we created the wings and married them to the computer shoe. In another case they had created a figure, but they couldn't find a way to make him swim, so we did that for them. All that we provide in those cases is the black outline and the computer then fills in the colours.

'There used to be a big divide between the traditional animators and video computer people, but now we are beginning to work together and it is creating good results.'

As with writers and directors of films and videos, there is a large pool of freelance talent in the animation world, most of whom work for the production houses. They are creative artists, and therefore have very distinctive styles of their own, some of which would be more suitable to certain projects than others, so it is worthwhile shopping around to see different people's work.

'How much you spend comes down to the design,' says Filmfair's Rob Dunbar, 'and there are a lot of exciting things which you can do if you have the budget.'

At Filmfair they specialise in making the three-dimensional series (most frequently seen in children's television programmes, where models and puppets move around with varying degrees of fluidity). The costs are similar to those of drawn animation, but it adds the possibility for including three-dimensional products in the action.

The other path into animation is through cheaper computer software systems, which are able to bring images on screen alive at relatively low cost. The quality is nowhere close to traditional cartoons, in fact it can be closer to the sort of animation which you saw on arcade games machines in the mid Eighties, but it can be a way of making dry information like charts and graphs come alive.

'Undoubtedly, computers will get better at it,' says Dunbar, 'and if someone comes up with a system which could automate the painting and tracing process, which is the least creative part of the operation, we could cut the costs enormously, and increase the speed with which we could produce results.'

INTO THE THIRD DIMENSION

It is in producing three-dimensional animations that computers provide the greatest scope for the future.

There are two distinct areas where this is developing. First there is computer aided design, where designers and engineers in areas like the car industry can actually work out how to build the best engines on screens, calculating things like stress factors. Then there is the graphic design area, where what matters is what the pictures look like and designers only need maths up to 'A' level standard to compete. It is in this latter field that most corporate work is likely to fall.

We have all seen it in the many tumbling logos and channel identification signs on broadcast television. We have also seen it in big-budget children's films and advertisements, although we may not have stopped to analyse what we are seeing.

When a teapot, in the middle of a live-action breakfast table, begins to walk around and contorts its face into different expressions, you can feel sure it has been 'built' on a computer.

What building on a computer means is that an object is drawn on a screen and is then 'digitised' in order to provide reference points. These points are entered into the computer for the making of a wire model on the screen, which can then be clad in whatever surface is required and turned into something solid. It can then be lit by computer, so that the shadows move with it, and animated and manoeuvred. Just as you can make a teapot come to life, you can do the same with a telephone, a car, a corporate logo, or you can build whole cities.

An area where the technique can be particularly useful is in the construction of lifelike models of buildings from plans.

'It is useful in any area where they traditionally have to build models,' says Linda Johnson, 3D consultant at Imagic. 'Architects, for instance, have to build models of their projects which can be cumbersome to move around. It is much easier for them to carry around a video, particularly if it is an international project involving transportation on planes.' A developer planning a shopping development, for instance, could make a lifelike three-dimensional film to show to potential buyers and leaseholders, as well as to local pressure groups, planning authorities, and the many other parties interested in seeing what the finished article will look like.

'We made a film for the Hong Kong government,' says Jay Williams, managing director of Digital Pictures, 'of their new airport development. They wanted to demonstrate to people exactly what it would look like, to give them confidence that it would happen. We "built" the whole thing, making it completely realistic. How realistic a model is will depend on the level of detail which you go into, and the amount of time you can afford to work on it.'

Once a model has been built the camera can travel around it, in the Hong Kong case exactly as if it were a real airport, going into buildings, landing on runways, flying over expressways and under bridges.

Digital Pictures did a similar job for the Department of Trade and Industry in Britain, creating a computer model of the British Pavilion to be built for the Barcelona Expo 92.

The same principles can be applied to models of anything which requires demonstration. You might, for instance, build a model of a car engine, and then take the camera – and therefore the viewer – for a realistic ride through the pistons and along any number of pipes and tubes.

Likewise you could build the inside of a human body and take the camera zooming through arteries and digestive systems to view cancerous growths, or anything else which the medical profession might want to demonstrate.

'Just as desktop publishing has revolutionised the way printed material can be presented,' says Williams, 'there will soon be equipment on the market which will make this sort of film-making more accessible to a wider range of people. We have already moved from an eight-foot long mainframe computer to workstation personal computers a few inches long, and there are new developments in hardware and software coming out every week.'

Chapter 16

Sponsoring Television

Sponsored television is a controversial subject in Britain. In America attitudes are very different, because sponsorship played such an important role in the development of the nation's television culture from the start. Soap operas after all – still one of the most popular of television formats – were brought to the public by sponsors like Procter and Gamble: hence the name.

In Europe the television establishments were even suspicious about the introduction of advertising in commercial breaks – the Swedish television networks started to allow advertising only in 1990, so Britain was by no means the last bastion to fall to the forces of Mammon.

Having accepted that commercials are a good way of funding programmes, the establishment (most notably the "independent establishment", if there can be such a thing, made up of the incumbent programme makers and broadcasters who have vested interests in maintaining the status quo), is now decrying the very idea of sponsored programmes as being a danger to everything that involves the creation of 'quality' television, not to mention the dangers which sponsors from the business community pose to their 'editorial integrity'.

'The critics of sponsorship complain about "editorial interference" from the sponsors,' explains Mike Morris, formerly head of Sponsorship, Licensing and Publishing at Yorkshire Television and now new business manager for Alan Pascoe Associates, a leading sponsorship consultancy. 'But to suggest that the programmes the 'independent establishment' makes are editorially independent is a falsehood. All sorts of things impinge on them. To start with, there are the beliefs and values of the people whon make the programmes, then there is the abundance or shortage of funds, and the pressures of the ratings systems.

'What these people are saying is that they are the only ones who are able to make valid decisions about what should and shouldn't be included in programme material. In my opinion, ITV people are not the only truthful people in the world.

'Sponsorship is potentially difficult and dangerous, and no one believes that it should be the sole source of finance for programming, but clients are mostly not interested in producing untruthful, biased programmes. They actually want to provide good quality, balanced and relevant programmes which will communicate successfully with their target markets. There is a general lack of understanding of what marketing is about in the television companies.'

The system of British commercial television has worked extremely effectively in the past, and may well be the ideal situation for providing a mix of programmes to suit the demands of the marketplace, but the changing market forces mean that the system is unlikely to generate enough money in the future to maintain its output of programmes.

'It is ideal to have large amounts of money arriving from a plurality of sources,' says Morris. 'Until now the selling of advertising time has produced much more money than the programme makers actually needed to make the programmes. It gave them a completely free hand, much freer than the BBC, who had one sponsor in the form of the government, but things are changing.

'With increased competition from other advertising media, a shrinking pool of available money and a steady escalation in production costs, it becomes obvious that other forms of funding are going to have to be found, and sponsorship is a good place to start.'

The difficulty is that many established broadcast television people take themselves very seriously indeed. They talk a great deal about 'art' and about 'editorial integrity' and other grand concepts, when most of them are merely crafts people turning out good quality programmes for a variety of different markets. They have also become used to being rich enough to do more or less as they please, within the restrictions which we have already discussed. They do not take kindly to the idea of having to go to outside sources for their financing, because that means they will have to share the power with a new set of people, people that many of them inherently despise. The original prejudices which I talked about in chapter 1 still exist in the hearts of many television people.

Many people in television are extraordinarily arrogant and have highly inflated ideas of their own importance. The less wordly they are, the greater these delusions seem to become, and the more they dislike and mistrust any suggestions of 'outsiders' being involved in the magic circle of television programme making.

Younger, less established programme makers, working hard to get their programmes made at all, will be less fussy about where the money comes from or whom they have to talk to in order to get it. They are also more likely to believe in the need for television airtime and programme-making opportunities to become more accessible to everyone and that the broadcasting and financing opportunities should be more flexible.

Someone who has not yet joined a television 'élite' will feel just as willing to talk to a high street bank, a supermarket chain, an insurance company or a petfood company about the need for finance as they will to an established television station. They may even find in these places fewer preconceptions and prejudices about what makes 'good television'.

Inevitably, this means that the new generation of programme makers will be more open to outside ideas, and will be more broad-minded about the ways in which television programmes are devised, financed and distributed. That change in mood, however, will not happen overnight.

Imagine, for instance, that you are a young programme maker with a script for a series about some characters travelling around Britain in a car, and the adventures they have along the way. It probably doesn't matter what sort of car it is, within certain boundaries of whether the characters are likely to be able to afford it. Suppose that you have raised nearly enough money to make the series, but you need another £50,000 or the whole project will collapse. Might it not make sense to go to a car company and suggest that in return for a topping-up on the budget you will use one of their cars in the series? Provided the programme is filmed in the way the creators envisioned, has the viewer been cheated in any way, bearing in mind that without the car company's money the programme probably wouldn't have been made at all?

From the point of view of the young programme maker, the constraints placed upon him or her by the motor company would probably be less onerous than those placed by a traditional television company, where executives have to keep their eyes on the ratings in order to protect their jobs.

CHANGES IN THE LAW

The laws governing the sponsorship of programmes are changing all the time, as the two sides lobby fiercely to get their points of view accepted. At the time of writing, companies are not allowed to sponsor a programme which is devised and made purely for broadcast television, but they *can* sponsor an event which will then be televised, as happens with the major sports events.

There are also chinks appearing in the system, whereby companies can put their 'labels' onto programmes, as Powergen has done with the weather reports.

There have, however, already been attempts by companies to move the game forward – most of them thwarted at some stage. Anchor Foods, for instance, sponsored a programme for children called *Go-Getters*, in conjunction with Yorkshire Television. It was a five-day treasure hunt held in the outdoors, which related to their advertising message. It was a successful format. Pedigree Petfoods produced a programme on pet care, which was then banned because it was decided that sponsors must have no direct interest in the subject matter of the programme which they are making. While it is possible to see why such an attitude has developed, it is a ludicrous approach and is bound, in the long term, to change.

Given the business they are in, who is in a better position to make a programme about pet care than a major petfood company? They have a company stacked with experts on the subject, and they have a vested interest in keeping the nation's pets happy and healthy, and in encouraging people to look after them correctly.

If the programmes were to be full of 'puffs' for the sponsoring company's products, then they would need to be regulated, but the television audiences are far from stupid and would soon switch off or disregard any programme which they felt to be biased.

A. C. Delco, the autoparts company, had an experience similar to Pedigree's after sponsoring a motorshow-style of programme. Yet these companies, I believe, will prove to have been the pioneers of a type of television sponsorship which is bound to grow in the long term.

At the time of writing, there is a draft code under consideration which suggests allowing all programmes to be sponsored apart from news and current affairs. It suggests that logos should be allowed to move or be animated, and that there should be no limit to the screen size. It proposes that programmes should be backed by either one or a number of sponsors, and credits should be permitted at the beginning, in and out of the breaks and at the end of the programmes. It thinks that credits should also be allowed to appear on trailers or promotional spots for the programmes concerned. Although verbal references to the client should not, they believe, be made during most programmes, they should be allowed in game shows.

TELEVISION COVERAGE AS AN ADD-ON BENEFIT

In most cases at the moment, however, companies involved in sponsorship on television come to it as a secondary consideration, having made an initial decision to sponsor an event for its own intrinsic values, with the televising of it as a bonus.

'The *Daily Telegraph* came to us with a brief to find a sponsorship programme which would help them to win a younger readership,' says Mike Morris, 'but which would not alienate their existing older readers. We came back with the suggestion of American football on Channel 4. It was an existing programme, but the paper was able to tie into it by producing American football supplements, and by using the presenter of the programme as a columnist in the paper.

'National Power did a similar deal with the World Cup via their advertising agency, and then they came to us to ask how they could make their sponsorship work for them. We developed it by approaching the *Sun* and sponsoring their World Cup coverage, with the corporate logo appearing on every page.

'Hewlett Packard, the technology company, came to us with the problem that they were technologists selling to technologists and they felt it was time that they started talking to a wider audience of senior managers. We suggested that they should sponsor the CBI's annual conference. What they did was supply communications equipment to the delegates for sending messages and for networking, which gave them the opportunity to contact each one individually to explain how the systems worked and to put across their corporate name. At the same time their name appeared alongside that of the CBI in all the television coverage. Again it was an example of television being an additional bonus to a sponsorship, when in this case the main target was the live audience at the conference.

'What sponsorship does is give real life values to paid-for advertising messages. It demonstrates a company's advertising messages. When Mars sponsored the London Marathon, for instance, it was a five-hour demonstration of how a Mars Bar helps you to "work, rest and play". It took the company into the community and demonstrated what they are saying in their advertising. The television cameras then carried that message to a wider audience.

'Sponsorship also gives companies an image of ubiquity, it gets them seen in areas they wouldn't otherwise reach and makes them seem an integral part of the nation's daily life. Instead of being limited to the supermarket shelf, they are seen as part of society at large.

'It can also be extremely useful for employee relations. If a company sponsors a sports programme, for instance, the athletes can visit factories and offices and meet people, and this can create a social atmosphere where one would not otherwise exist. Again, it is a question of bringing real life values to bear.'

In the long term, there is no reason why the same sort of sponsorship deals shouldn't work with every programme. Tesco, for instance, could sponsor *Coronation Street*, having access to the stars for store openings and staff gatherings: Fisons might sponsor *Emmerdale Farm*. In some cases it could be that the sponsor's name would actually bring a lustre to the programme which it wouldn't otherwise have, as the *Financial Times* has done with its satellite links. Experts from the famous arts houses, for instance, already add credibility to the *Antiques Road Show*, so why shouldn't engineering experts at Ford give credibility to a programme on car maintenance, or experts from the Prudential appear on personal finance programmes?

Traditionally the ideas are hatched in the television companies and the experts are then approached in an arbitrary fashion to help with the presentation or research of the show. There is no reason, ultimately, why the concepts of the shows could not be hatched within industrial or service companies with a particular expertise, and why the television producers should not then be approached to find ways and means of making and distributing the resulting programmes.

LABELLING OPPORTUNITIES

The pure 'labelling' of programmes carries fewer benefits for the sponsors, and is probably of less value than simply buying advertising slots around the chosen programme. With labelling, you can buy a credit at the front of the programme and another at the end, and perhaps something at each end of the commercial break, but it will be no more than a simple name check. At least with a traditional advertising spot, a client is able to put across whatever marketing message they wish.

At the moment the restrictions on what can be done with corporate logos (they basically have to stand still, even if all the other credits are animated) can make a sponsoring company look old-fashioned. That, however, is one of the things which is likely to change in the next few years.

There are some startling examples of what can be achieved with animation from companies like Pepsi. Kellogs attempted to use animation to create a series of children's programmes called *The Magic Mirror*, but had to withdraw the results in the face of accusations of putting across subliminal messages when they wove their 'K' into the animation. The area of children's television is particularly sensitive, with no company wanting to be accused of using unethical practices.

From all this chaos and change, however, some clear messages are emerging. With the diversity of television stations opening up, there is bound to be an increased need for television programming of a sufficiently high standard to satisfy viewing requirements. Companies which contain large amounts of expertise on any particular subject, and which have a marketing need to talk to wide audiences, are in a particularly strong position to provide these programmes, although they may not be able to get exactly the outlets they want for them right away.

CREATING YOUR OWN PROGRAMMES

One way to approach the problem is the traditional route of choosing some sort of event to sponsor and then finding ways of getting that event televised.

The next option is to find some way of creating your own programmes, with an internal use for them guaranteed, and then looking for other outlets through the many burgeoning satellite, cable and independent channels.

A petfood company, for instance, could go ahead making pet care programmes which could be distributed as videos to pet shops, shown at animal-related events and promoted as products in the same way as books have been for years. If the programmes are of a sufficiently high standard they will, eventually, find their way onto more widespread systems of exposure. Nothing is guaranteed, of course, but the companies who start investing early in these sorts of projects will be the ones who have the material available as the television stations' appetite for programmes grows.

There are going to be so many 'special interest' channels of television before long, aimed at business people, doctors, the high technology industry, housewives, and hundreds of other niche sectors, that the demand for good material will be massive.

Creating Broadcast-Standard Programmes from PR Material

Just as the existing public relations industry supplies a vast proportion of the printed material which appears as editorial in specialist magazines and newspapers, television will have to look to the business world for help in supplying material which interests their viewers.

It may be that the seeds of a broadcast television programme already exist within a company, and merely need ferreting out and packaging by the public relations department.

When the City of London commissioned a documentary about itself, the orginal intention was to use it for their own public relations. The fact that World Wide Pictures, who made the film, were also able to sell it to the BBC meant that the PR value was increased, while the fee helped to cut the cost of production.

For many years the Unipart Company have been producing a glamour calendar for distribution to the trade. Each year photographer Patrick Lichfield heads off to another exotic location with a bunch of models and produces a new set of pictures. The PR value of such a project is enormous, but it reached its heights when they managed to persuade the BBC to run an hour-long documentary about the making of the calendar in Kenya, with Clive James performing a sardonic voice-over. The company made the programme and presented it to the BBC free of charge. It was shown at peak viewing time. The idea worked so well that in the following five years they were able to do a similar deal with ITV companies, giving them films of calendars being shot in Bali, Le Mans, Death Valley, Moscow and New York.

Unipart is a company which has been particularly imaginative in its use of television at all levels. When their public relations director Patrick Fitzgibbon heard that the makers of *Howards Way* were writing the sport of powerboating into the TV series, he offered them the use of the Unipart boat, which became a regular in the series. The company also sponsored the Unipart International Darts championship on the BBC for about eight years, sometimes getting as much as twenty-two hours of coverage a week.

'We have three hot air ballons as well,' says Fitzgibbon, 'and because they are such nice things visually, we have managed to arrange a lot of television coverage for them. We taught one TV presenter how to pilot a balloon, which resulted in a half hour documentary, and we took another camera crew over to Leningrad on a flight.'

'The Unipart calendar programmes illustrate my point,' says Mike Morris of Alan Pascoe Associates. 'They were quality television programmes, because Unipart had a vested interest in creating something which would be good for their image and which would attract a lot of viewers. If it hadn't been funny and interesting, it would not have been screened.'

There must be opportunities like this in every major company, and if they don't already exist, there must be ways of making them happen, as Richard Branson does so skilfully with his various escapades around the world, accompanied by the Virgin Group's logo.

The world is hungry for good television programmes and grows hungrier every year. Companies will be supplying that need in one way or another.

'There is going to be a massive change in the way that all television is financed in the future,' says Ray Townsend of World Wide Pictures, 'due to legal and technological developments. The production houses which will benefit, we believe, are the ones which have experience in the corporate and the broadcast sectors, because they know how to satisfy corporate client requirements and how to make programmes to broadcast standards. I firmly believe that in a free market situation sponsorship need not have any detrimental effect on the quality of programming.'

Chapter 17

The Future is Here

People have been talking about the coming of cable television in Britain for so long that those who still haven't got one within a few miles of their home have begun to wonder if it exists at all.

It does exist, and it has taken a great deal longer that anyone predicted, but during the Nineties it should begin to take a stronger hold. If enough investment goes into it, eventually cable could turn the satellite dish into a dinosaur of the early communication revolution, and we will all be falling about laughing at the memory of how we stuck those ludicrous things all over our houses.

The cable system works rather like the piping of gas or the construction of telephone lines. The cables are laid in certain areas, and if one runs close to your house or office you can tap into it and gain access to the various channels and programmes on offer. Like gas, there will be some rural areas which will probably never be reached by cables, but within a few years even they will be able to receive the channels via microwave, which will involve a satellite dish about the size of a saucer. (It is the televisual equivalent of Calor Gas, really, the disadvantage being that it will work for one-way only, not for two-way communication.)

Cable isn't a rival to satellite, it is merely a different means of distribution for programmes, like the difference between traditional telephone lines and mobile phones; the services are the same, the messages simply travel a different way. All that happens is that the satellites beam their programmes to the cable stations, which then feed them down the lines to their customers along with the range of other programmes and services.

The advantage in the long run will be that you will be able to receive anything you care to subscribe to without having to load down your roof with satellite dishes.

At the time of writing, there are about 700,000 homes connected to cable out of twice that many which have a cable passing them. (There is a total of 22 million homes in Britain.) But the Cable Authority, which was set up to dispense operating licences, predicts that by the middle of the Nineties about 70 per cent of the population will have access to the service.

The opportunities for business television will begin to appear once all the major cities are linked. In America it is used for closed user operations like the Financial News network, which is a concept that could work in Britian, (probably the *Financial Times* would be the ones to organise it). The Cable News Network is already carrying world news services around Europe.

Any company which does a lot of business with another country can use cable to receive that country's news broadcasts in order to keep up with the domestic scene in their marketplace. Just as they might buy a German or French paper to make sure they were keeping up with events, they can tune in to their television programmes, whether it is Europe, Russia or the Middle East.

COMPETING WITH THE TELEPHONE SERVICES

The great selling point for cable is that it could provide telephone services, if the existing laws can be adjusted, and dedicated local stations. It also provides video conferencing opportunities because eventually, once everyone is wired up, you should be able to plug in a camera anywhere and broadcast to the location of your choice.

The fact that the cable television operators want to supply telephone services down their lines also opens up the possibility that British Telecom might one day be allowed to send television programmes down its lines.

HIGH-DEFINITION TELEVISION

At the time of writing, HDTV for broadcast television is still something of an unknown quantity, because of problems of establishing a worldwide standard. In a business context, however, it is already arousing interest.

It is, as its name suggests, a system which creates far better, clearer and more detailed pictures than traditional television or video. This solves one of the problems with television pictures which when thrown up onto a big screen become fuzzy or lined and thus inappropriate for some detailed applications or for major presentations.

The Conference Market

Many conferences and exhibitions are forced to use tapes and slides to achieve the right degree of clarity and quality on the big screen. HDTV changes all that, being able to compete with slides on quality while producing a moving image. It therefore has a number of uses.

In any sort of presentation situation, HDTV demands the attention of the audience, using larger screens and producing more realistic pictures. Whereas it is easy for a viewer's attention to wander from a traditional television picture, the HDTV screen holds their eyes and minds in much the same way as a cinema screen would. There are also some huge technical possibilities.

The Product Design Market

Ford, for instance, has recently announced a £10m investment in an HDTV-based design system from Sony, which underlines the way in which computer and HDTV technology are moving closer together, and increases the likelihood that Sony will be at the front of any race to get HDTV to a mass market.

BUSINESS TELEVISION LEADS THE WAY

The high development costs of HDTV has forced companies like Sony to seek business applications to provide the necessary finance, while they find ways of getting into the broadcast market. Sony is launching a £65,000 HDTV cassette system aimed at the corporate market.

In the past it has been the broadcasters who led the way by having the newest technology, with everyone else following on behind, but the boot is now shifting onto the other foot. Although broadcast television will still be the most influential sector of the market, and 'film' the most developmental, it is the business television market that will pay for the development of HDTV.

The Japanese system is also being adopted for industrial, printing, medical and military applications. It is particularly useful for anything to do with design, historical analysis or long-range surveillance. It can also be used to produce more realistic experiences for things like flight simulators.

Two of the first Sony HDTV systems were sold to Creative Technology for use in the conference market. 'Basically HDTV gives us video which looks like 35mm film in quality,' says their managing director Graham Andrews. 'It gives us the flexibility of video, combined with the quality of slides, because it has 1125 lines instead of the traditional 625. It works on a wide-screen format, so it provides a more cinematic experience for the audience. It is still a very expensive business to get into, but in the long term it is bound to become standard for big-scale productions.'

Ford wanted HDTV projectors which were capable of giving life-size, accurate pictures at least five metres wide. It was so pleased with the use of the HDTV system in its Cologne engineering design department and the British interior design unit at Dunton that it is, at the time of writing, planning to install such systems in another twelve design centres. It uses fibre optic cable to transmit pictures between Cologne and Dunton, not wishing to use satellites for security reasons.

The company believes that HDTV can cut its design cycle on a new car from 64 months to 48, reducing the number of clay models for each design it makes from over 40 to just two or three.

Fiat and BMW have also installed in-house HDTV production units for commercials and promotional materials, and Ford is planning to produce HDTV animations which allow the viewer to roam the interior and exterior of the car.

WHAT NEXT?

All the ingredients for how television will affect our immediate future are in place.

There are still legal matters to be settled in some areas, but the hardest tasks for the industry as a whole are now the actual setting up of the infrastructures which will make its services available to a wider audience, and then of educating the world to an awareness of the benefits of the products.

We already know how to put personal video conferencing facilities on everyone's desks.

We already have personal computers which allow their users to create their own video programmes.

We can create private networks for anyone who is willing to pay for them.

We can create interactive programmes which virtually recreate reality to our own design.

Animators can now create three-dimensional films which will create previously impossible experiences for the viewer.

We can build screens of virtually any size, and beam pictures down from anywhere on earth.

Televisions can now hold conversations with us.

They can do business with us.

They can teach us, and show us ourselves.

They can bring us experiences which would never have been possible any other way.

They can transport us across time zones.

All these things are happening at the leading edges of the industry, but none of them can become commonplace overnight.

Just as the computing revolution took twenty years to establish itself, the business television industry has a long, hard educational job ahead of it.

People need to feel comfortable with the technology. They need to have seen it in action to know what it is capable of doing. They need to grow accustomed to using it and confident about appearing on the screens.

Then they will find it hard to remember how they ever managed without it.

MAKING IT ACCESSIBLE TO EVERYONE

Some television professionals claim that the potential of the market is limited, because there aren't enough people sufficiently talented to communicate effectively through television screens.

In the early days of the automotive industry, a survey suggested that the potential market for motor cars was severely restricted because there was a limit to the number of professional chauffeurs in the world.

There is still a limit to the number of chauffeurs in the world, and there are even fewer skilled racing drivers, but the motor car industry has succeeded in changing all our lives almost beyond recognition in just two generations.

Now it is the turn of television.

Index

Abbey National, 121, 124
Alan Pascoe Associates, 144, 151
Alliance & Leicester, 126
Allied Dunbar, 120, 122
American Express, 120, 122
Anchor Foods, 147
Andrews, Graham, 26, 75, 156
Antiques Roadshow, 149
Apple Computers, 75
Aristotle, 99
Artificial Intelligence Satellite Symposium, 81
Austin Rover, 120, 122

B&Q, 120
BBC, 6, 7, 13, 103, 111, 145, 151
BMW, 26, 51, 156
BOC, 131
BP, 34, 35, 62, 66
BSB Datavision, 15, 48–50, 58
Bank of England, 121
Bank of Scotland, 121
Bankers Trust Co, 121
Barclays, 30
Batiste, Stephen, 43
Blackrod, 8, 23, 24, 36, 43, 54, 65, 134
Blanchard, Kenneth, 103
Bowman, Lee, 117–119
Branson, Richard, 152
Bratt, Chris, 35
Bright Star, 47, 80, 81, 84, 85
British Aerospace, 29, 53, 71
British Airways, 122
British Audio Visual Society, 124
British Rail, 31, 43
British Steel, 120
British Telecom, 31, 42, 70, 71, 79, 82, 83, 120, 126
Brown, Andrew, 31, 32
Brunel University, 100
Burgess, Colin, 100
Business Daily, 14
Business Report, 15
Businessweek, 14
Byron, 99

CBI, 148
Cable Authority, 153
Cable News Network (CNN), 14, 15, 154
Castles in the Air, 91
Carters PR, 79
Channel 5, 30
Channel Four, 14, 148
Chrysalis Interactive Services, 126
Citicorp, 121
Clarke, Bob, 26, 51, 52
Cleese, John, 9, 68, 103, 132
Cole, George, 135
Cole, Peter, 38, 104, 108
Coles, Neil, 49, 50, 58, 59
Commercial Presentations, 43
Computerland, 47
Co-operative Wholesale Society, 124
Coronation Street, 149
Corporate Vision Ltd, 111
Creative Technology Ltd, 26, 29, 70, 75, 155

DHL, 52
Daily Telegraph, 148
Darwin, Charles, 99
Datawize, 85, 86
da Vinci, Leonardo, 99
Delco, A. C., 147
Dept of Trade and Industry, 142
Dickens, Charles, 99

Digital Computers, 60
Digital Pictures, 142
Disney World's Space Mountain, 128
Disney, Walt, 88
Disney Studios, 139
Domino Pizzas, 47
Duke, Roy, 14
Dunbar, Rob, 139, 141

Edmonds, Noel, 132
Emmerdale Farm, 149
Euromoney Publications, 105
Europace, 100
European Business Channel, 15
European Commission, 111

Fate of the Forests, 9
Federal Express, 47, 126
Fiat, 156
Filmfair Animation, 139, 141
Financial Animation, 139, 141
Financial i, 105, 106
Financial Times, 13–15, 49, 149, 154
Fisons, 149
Fitzgibbon, Patrick, 30, 151
Fonda, Jane, 104
Food and Famine, 9
Ford, 46, 80, 83, 121, 149, 156
Futura, 40, 41, 91
Future Shock, 59

GHA, 20, 21, 28, 64, 65, 110, 135, 138
Gastrolink, 81, 82
General Motors, 46
General Portfolio Assurance, 32
Glaxo, 81
Go Getters, 147
Gorbachev, Mikhail, 39
Gower, 103
Grapevine, 30
Great Bridges of the World, 8
Gye, Roddy, 20, 21, 28, 65, 110, 135, 138

Harvard Business School, 103
Harvey-Jones, Sir John, 103
Hewlett Packard, 100, 148
Higginson, Graham, 23–25, 54
Hobday, Peter, 49, 132
Horizon, 8
Howards Way, 151

IBM, 47, 71, 86, 100, 101, 124, 128
ICI, 103
ICL, 71
ITV, 13, 144
Imagic, 142
In Search of Excellence, 103
Intercontinental Hotel, 79
International Visual Communications Association (IVCA), 10, 11, 19, 23
Jackson, Michael, 84
Jaguar, 120, 125, 126
James, Clive, 151
Japan Airlines, 64
Jason, David, 132
Johnson, Linda, 142

Kelloggs, 149, 150
Kingstree Group, 117, 118

Leyland Daf, 66
London Scientific Films, 32
Lichfield, Patrick, 151
Limehouse Studios, 131
Littlewoods, 120
Lloyds Bank, 106
Lodge, Mike, 140
Longman, 103

Macintosh computers, 75
The Magic Mirror, 149
Marks and Spencer, 42, 64, 105
Mars, 148
Martin, James, 101, 102, 104
Maxwell Group, 79, 83
McCann Erickson, 84
McCluhan, Marshall, 102
Melrose Training, 38, 103, 104, 108, 109
Medicine Now, 82
MetroVideo, 84
Merck Sharp and Dohme, 82
Merrill Lynch, 47, 54
Mersey Docks and Harbour Co, 62
Michie, Maxwell, 9, 33, 60, 98, 111, 140

Midland Bank, 122
Ministry of Defence, 122
The Money Programme, 6, 7, 14, 40, 111, 133
Morris, Mike, 144, 145, 148, 151
Murray, Hugh, 104

National Exhibition Centre (NEC), 70, 72
National Interactive Video Centre, 129
National Power, 148
National Westminster Bank, 49, 58
New Media Sales, 14
News at Ten, 15, 75
Newsnight, 39, 75
Newsweek, 14

Olympia Conference Centre, 84
One Minute Manager, 103
Open Air, 134
Orwell, George, 53, 57

Panorama, 39, 111, 133
Pauffley & Co, 131
Pedigree Petfoods, 147
Pentony–Woolwich, Martin, 21, 70, 71, 133, 134
Pepsi, 84, 149
Peters, Tom, 68, 101, 103, 104
Philips, 90
Pipeline, 62
Plato, 99
Playback, 103
Porter, Michael, 68, 103
Powergen, 70, 147
Price Waterhouse, 49, 106, 120, 121
Procter & Gamble, 144
Prudential, 149

Question Time, 103

RAC, 30
Radio Rentals, 50, 55, 135
Rank Training, 103
Reuters, 80
Rival World, The, 9
Rhys Jones, Griff, 103, 132
Rodd, Michael, 7, 36, 43, 44, 65, 134
Roles, Rosita, 40, 41
Ross, Nick, 132
Rothmans, 8
Royal Bank of Scotland, 83
Russell, Bertrand, 99
Satellite Management International (SMI), 53, 81, 82
Savoy Hotel, 72
Scott, Patrick, 15
Shakespeare, 99
Shell, 9, 30, 33, 34, 39, 60, 98, 111, 120, 122, 140
Sheppard, Roy, 116, 134, 135
Singapore Airlines, 65
Singleton, Valerie, 132
Sky Television, 16, 48
Smith, Mel, 103, 132
Smithsonian Museum, 102
Sony, 155
Sony Walkman, 128
Spectrum Communications, 21, 70, 71, 133
Sportsnight, 133
Sullivan, Colin, 105
Superchannel, 14, 15

3M, 123
Telesensor, 94
Tesco, 88, 149
Texas Instruments, 46, 53, 81
Thompson, Peter, 53, 82
Time, 14
Today, 111
Toffler, Alvin, 59
Tomorrow's World, 8, 133, 134
Too Fussy for Words, 42
Top Gear, 133
Townsend, Ray, 8, 9, 19, 22, 34, 36, 62, 66–68, 152
Toyota, 85
Trilion Group, 91
Troubleshooter, 103

Unipart, 30, 121, 151
Universal Video Communicator, 85

VAG, 32
Valin Pollen, 79
Video Arts, 9, 103, 107, 132

Vidiwall, 90
Visage, 26, 29, 51

Wang, 46, 47
Watson, Alan, 6, 7, 111–113
Webster, Mark, 86
Wellcome, 30, 31, 80
Whitehead, Frances, 81
Williams, Jay, 142, 143
Wimbledon, 89
Wogan, Terry, 132
Woodward, William, 132
World Business Tonight, 14
World Wide Pictures, 8, 19, 22, 34, 36, 62, 66, 67, 151, 152
Wyvern Crest, 103, 104
Wyvern Television, 50

Xerox, 47

Yorkshire Television, 144, 147
Young & Rubicam, 80

The Mercury titles on the pages that follow may also be of interest. All Mercury books are available from booksellers or, in case of difficulty, from:

Mercury Books
Gold Arrow Publications Ltd
862 Garratt Lane
London SW17 0NB

Further details and the complete catalogue of Mercury business books are also available from the above address.

WINNING WAYS

James Pilditch

How has Alan Sugar grown from selling car aerials off the back of a truck to owning one of the most successful companies in the world? How did Anita Roddick, a 33-year-old housewife with two children, become Business Woman of the Year? What does a firm in the Staffordshire countryside (JCB) do to be market leader in 56 countries? What gives Sony and Sharp and Canon and Minolta their edge with customers? Why are IBM and Philips and Courtaulds and Honda always in the lead?

Winning Ways is written for all who want their businesses to do better – directors, engineers, designers, marketing, R & D and financial people – and for educators who, as the book suggests, have a larger role to play.

'James Pilditch has produced a book which should be compulsory reading for managers. Britain has the conditions now which enable its manufacturing industry to survive. What we need is many more winning products. James Pilditch has shown the way to get them.'
Sir Simon Hornby, chairman W H Smith, chairman Design Council

'. . . very nicely instructive, full of strong examples and focused reasoning. I especially like the fast-paced urgency that moves the reader along as he constantly gets helpful new advice and insights. The book should be a winner.'
Theodore Levitt, Editor, *Harvard Business Review*

'In this wonderfully clear book it becomes obvious that the author is a man of vision . . .'
Robert Heller

ISBN 1 85252 042 6 (paperback) £7.99 net

TURNING PRACTICAL COMMUNICATION INTO BUSINESS POWER

Bernard Katz

Good communication is a direct pathway to the ingredients of business power. Effective communicators exert influence. They establish authority, promote understanding, and they persuade. Good communications is as essential to the corporate body as it is to the individual manager.

In his latest book Bernard Katz provides the reader with a range of practical communication skills. They apply to the three major areas of business:

- The company communicating with its customers and public
- Management and staff communicating with each other
- The individual manager and executive who must make personal presentations, conduct meetings, prepare reports, promote a product of service, and communicate professionally on the phone.

Getting better at communicating is easy. There are examples, with check lists and rules, and dire warnings of what not to do. Guidance is also provided for the selection of audio-visual back-up to the communication message. And there is a section, with training guidelines and examples, for those whose task is to train others to communicate more effectively.

Turning Practical Communication Into Business Power is essential reading for all concerned with promoting and manipulating business understanding. It pays particular attention to the private apprehensions of those learning to overcome communication pitfalls.

'Highly readable and packed with invaluable advice for managers at all levels.'
Roy Sheppard, LBC

ISBN 1 85251 076 5 (hardback) £12.95 net

MARKETING TOOLKIT

Nick Robinson

Nick Robinson's *Marketing Toolkit* is an encyclopaedia of more than 1000 fresh, tested marketing ideas to improve your business success.

This intensely practical book illustrates a wealth of ingenious new ways for your organisation to create profitable sales. It details advertising approaches – including displays, classified, loose inserts, directory, product cards and local radio. It reveals direct marketing methods, including ways to boost direct mail and direct response advertisements. And it gives specific step-by-step guidance in planning, budgeting and executing effective sales lead generation programmes.

All organisations can profit from these ready-to-go plans, regardless of their type of market or size of budget. And any reader who has the task of making marketing succeed – whether you direct other people or do the job yourself – will benefit from this down-to-earth advice, based on the author's 20 years' personal experience of what actually works.

'If it's marketing ideas you want, don't look further.'
John Winkler, Winkler Marketing

ISBN 1 85251 002 1 (hardback) £12.95 nct